Science for Sustainable Societies

Scope of the Series

This series aims to provide timely coverage of results of research conducted in accordance with the principles of sustainability science to address impediments to achieving sustainable societies – that is, societies that are low carbon emitters, that live in harmony with nature, and that promote the recycling and re-use of natural resources. Books in the series also address innovative means of advancing sustainability science itself in the development of both research and education models.

The overall goal of the series is to contribute to the development of sustainability science and to its promotion at research institutions worldwide, with a view to furthering knowledge and overcoming the limitations of traditional discipline-based research to address complex problems that afflict humanity and now seem intractable.

Books published in this series will be solicited from scholars working across academic disciplines to address challenges to sustainable development in all areas of human endeavors.

This is an official book series of the Integrated Research System for Sustainability Science (IR3S) of the University of Tokyo.

More information about this series at http://www.springer.com/series/11884

Takashi Mino • Shogo Kudo
Editors

Framing in Sustainability Science

Theoretical and Practical Approaches

 Springer Open

Editors
Takashi Mino
Graduate Program in Sustainability
Science-Global Leadership Initiative
Graduate School of Frontier Sciences
The University of Tokyo
Kashiwa, Chiba, Japan

Shogo Kudo
Graduate Program in Sustainability
Science-Global Leadership Initiative
Graduate School of Frontier Sciences
The University of Tokyo
Kashiwa, Chiba, Japan

Department of Socio-Cultural
Environmental Studies, Graduate School
of Frontier Sciences
The University of Tokyo
Kashiwa, Chiba, Japan

ISSN 2197-7348   ISSN 2197-7356   (electronic)
Science for Sustainable Societies
ISBN 978-981-13-9060-9   ISBN 978-981-13-9061-6   (eBook)
https://doi.org/10.1007/978-981-13-9061-6

This Springer imprint is published by the registered company Springer Nature Singapore Pte Ltd.
The registered company address is: 152 Beach Road, #21-01/04 Gateway East, Singapore 189721,
Singapore

Preface

Sustainability science emerged in the early 2000s as a new academic field to address sustainability issues through problem-driven and inter- and transdisciplinary approaches. The field sets its primary purposes in understanding the complex interactions between the ecological system and human society, in elucidating norms and values related to sustainability, and in proposing new technological or social approaches that move entire societies toward sustainability.

Framing is an essential process in sustainability science. This is because sustainability is fundamentally a normative concept: how people view the world influences what topics should be considered as problems and how such problems should be framed in the sustainability manifestation. Framing explains how people perceive and interpret particular topics or events based on the social norms, values, and assumptions that they apply in each situation. In reality, multiple framings by different groups of people always exist in a society because of the different understandings of reality; hence, diverse interpretations of situations always exist. Reflecting such multiplicity in people's framings, experts who can facilitate collaborations among different social actors to lead transformations towards sustainable society are needed.

Scholars in sustainability science also hold different understandings of reality and different framings for addressing sustainability issues. Furthermore, sustainability science stresses co-creation of knowledge and co-design of actions for sustainability between academic and various social actors; this implies a convergence of a greater degree of differences in framing. To perform inter- and transdisciplinary approaches effectively, acknowledging the presence of multiple framings and learning ways to create synergy among people who have different framings are necessary.

This book attempts to introduce both conceptual and practical framings applied in their respective fields by inviting authors from two graduate programs that are offering sustainability science degree courses, namely, Graduate Program in Sustainability Science-Global Leadership Initiative (GPSS-GLI) of The University of Tokyo and Lund University Centre for Sustainability Studies (LUCSUS) of Lund

University. By doing so, this book aims at providing an overall picture of diverse framings applied in sustainability research and education and giving theoretical as well as practical bases of framing in sustainability science to those who are motivated to guide our society to sustainability, thus becoming sustainability experts.

Kashiwa, Japan   Takashi Mino
                                                                                              Shogo Kudo

Contents

Part I
Theoretical Approaches
to Sustainability Issues

Chapter 1
Framing in Sustainability Science

Shogo Kudo and Takashi Mino

Abstract This chapter discusses multiple understanding of sustainability by examining the process to identify what must be framed as sustainability challenges. The chapter first provides a summary of past development of sustainability science as a new interdisciplinary filed that sets its primary purposes in understanding complex human-nature system and academic knowledge contribution to the pursuit of sustainable development. To elaborate some of the educational features of sustainability science, brief history and curriculum design of Graduate Program in Sustainability Science – Global Leadership Initiative (GPSS-GLI) of The University of Tokyo is introduced. One central question in sustainability science is "what to frame as sustainability challenges?". The chapter employs the concept of framing to examine what topics to be included and how they should be discussed in sustainability science. Framing explains how people perceive and interpret particular topics or events with the social norms, values, and assumptions that they apply in all situations. Being self-aware about what type of framing is used when discussing particular sustainability challenge is critically important. At the last, the chapter proposes a conceptual framework that includes holistic treatment, resilience, and trans-boundary thinking to depict multi-level dynamics of sustainability challenges. This framework serves as a guideline to (i) analyze the complexity of sustainability issues through multiple framings, (ii) apply holistic treatment and trans-boundary thinking in the process of developing action plans, and (iii) evaluate the proposed

S. Kudo (✉)
Graduate Program in Sustainability Science-Global Leadership Initiative,
Graduate School of Frontier Sciences, The University of Tokyo, Kashiwa, Chiba, Japan
e-mail: kudo@edu.k.u-tokyo.ac.jp

T. Mino
Graduate Program in Sustainability Science-Global Leadership Initiative,
Graduate School of Frontier Sciences, The University of Tokyo, Kashiwa, Chiba, Japan

Department of Socio-Cultural Environmental Studies, Graduate School of Frontier Sciences,
The University of Tokyo, Kashiwa, Chiba, Japan
e-mail: mino@k.u-tokyo.ac.jp

© The Author(s) 2020
T. Mino, S. Kudo (eds.), *Framing in Sustainability Science*,
Science for Sustainable Societies, https://doi.org/10.1007/978-981-13-9061-6_1

actions from the perspectives of both top-down approaches and bottom-up approaches. The authors believe that sustainability experts must be trained with knowledge and skills to utilize this framework in sustainability research and action projects.

Keywords Framing · Sustainability science · Holistic treatment · Resilience · Trans-boundary thinking

1.1 Emergence of Sustainability Science

The idea of sustainable development—fulfilling and enhancing human well-being while sustaining the life-support system of the earth—was introduced globally by the report of the World Commission on Environment and Development, *Our Common Future*, in 1987 (WCED 1987). The United Nations Conference on Environment and Development in 1992, also known as the Rio Summit, reached agreement on the commitment of academia to actively engage in addressing development and environmental problems (UNESCO 2000; Lubchenco 1998). Supported by these international conventions, the idea of sustainable development was recognized as the main direction of development for the twenty-first century.

What is required to create a transition of human society towards sustainable development is a fundamental understanding of the relationships between humans and nature, and of the methods to transform such knowledge into actions (Phillips 2010). In response to the sustainable development discourse, sustainability science has emerged as a new academic field that sets its primary aim as advancing the understanding of the complex interactions between social systems and natural systems (Clark 2007; Kates 2001; Martens 2006; Ostrom et al. 2007; Swart et al. 2004). Sustainability science aims at understanding "how social change shapes the environment and how environmental change shapes society" (Clark and Dickson 2003). Komiyama and Takeuchi (2006) introduced a similar perspective by explaining sustainability science as a field of comprehensive studies on the multi-scale and complex interactions among three sub-systems: global, social, and human systems (Komiyama and Takeuchi 2006).

Reflecting the focus on the interactions between social systems and natural systems, sustainability science addresses challenges that include complex structures within themselves. Being complex, in this context, refers to the presence of dynamic system-subsystem relationships in a human-nature system. These interactions exist across multiple spaces, times, and scales from local to global; and each subsystem has its own particular qualities and properties (Rosen 2005; Satanarachchi and Mino 2014). Systems dynamics perspectives play an important role in sustainability science illustrating such complex interactions in system-subsystem relationships (Fiksel 2003; Kinzig et al. 2006; Morse 2010; Vries 2013). Complex challenges in sustainability science are exemplified by climate change, biodiversity loss, deforestation, rapid urbanization, poverty and hunger, epidemics, and natural disaster management to name a few (Jerneck et al. 2011; Leeuw et al. 2012; Rosen 2005). These sustainability challenges are constantly changing over time; hence, actors must often develop temporal approaches to the problems simultaneously in analyzing the problem structures when addressing sustainability challenges (Hiramatsu 2012; Komiyama and Takeuchi 2006; Sterman 2012).

To analyze the complexity of sustainability challenges and guide the current human-nature system to pathways to a sustainable state, sustainability science must differ in its structure and approaches from conventional science based on a reductionist perspective, and must also incorporate adaptive management, problem-based and action- oriented perspective, and social learning approaches (Clark 2007; Kates 2001; Weinstein and Turner 2010). Reflecting these characteristics, sustainability science is an interdisciplinary field in which different bodies of academic knowledge are integrated. Furthermore, sustainability science is presented as a transdisciplinary field that combines knowledge not only within academia, but also with various social actors (Kajikawa 2008; Lang et al. 2012). This transdisciplinary orientation implies designing a transformational change of the current situation to lead our society to a sustainable pathway (Chapin et al. 2011; Leeuw et al. 2012; Wiek et al. 2012). Such an idea of producing collaborative knowledge and implementing action is not limited to sustainability science; however, some experts in sustainability science explicitly emphasize its nature as being *transformational science* (Wiek and Lang 2016). Pre-1987 literature illustrates the emergence of sustainability science both quantitatively and qualitatively.

Some bibliometric studies that analyzed the publication and co-authorship in the field of sustainability research (Bettencourt and Kaur 2011; Kajikawa 2008; Kajikawa et al. 2014; Kates 2011; Schoolman et al. 2012) found a steady increase in sustainability research not only by the number of publications, but also by the increase in thematic coverage. One major research theme identified is *resilience* (Kajikawa et al. 2014). In the studies that examined sustainability science more qualitatively, sustainability science can be subdivided into science *for* sustainability and science *of* sustainability (Spangenberg 2011). Science *for* sustainability provides more technical approaches to offer possible solutions to sustainability challenges, and implies a basic scientific method based on problem-based and interdisciplinary approach. The science *of* sustainability, however, aims to develop a conceptual and methodological discussion of sustainability, the observation of which "can be understood as a new step in the evolution of science" (Spangenberg 2011).

While some scholars treat sustainability science as an independent discipline, others argue that sustainability science is rather a discipline that accommodates diverse interactions among different academic disciplines (Clark and Dickson 2003). Shahadu (2016) claims sustainability science is an "umbrella science" that fills the gaps among different research traditions based on different ontologies and epistemologies (Shahadu 2016). Acknowledging such multiple understandings of the concept of sustainability and facilitating interdisciplinary communication are necessary steps for sustainability science to recognize its pluralistic nature (Olsson et al. 2015).

Despite the evolvement of sustainability science as a new academic field, some major challenges remain in its conceptual and methodological developments as well as in its establishment of institutional structures that fit its inter- and trans-disciplinary orientation (Yarime 2013). Further development of the field is expected in the direction of realizing multiple understandings of the concept of sustainability and manifesting such ideas into institutional arrangements. This chapter aims to contribute to enhancing multiple understandings of sustainability by examining the process to identify what must be framed as sustainability challenges. To conclude the chapter, the authors propose their conceptual framework of key elements for visualizing transformation to a sustainable society. The next section introduces the sustainability science program at The University of Tokyo to present some of the key features of sustainability science education.

1.2 Educational Challenge in Sustainability Science at UTokyo

The Graduate Program in Sustainability Science–Global Leadership Initiative (GPSS-GLI) (http://www.sustainability.k.u-tokyo.ac.jp) is a combined Master's and Doctoral degree program based at The University of Tokyo (UTokyo). The program offers an interdisciplinary education over five years – generally two years for a Master's degree, and three years for a Doctoral degree – and aims at fostering leaders for developing sustainable societies. The integrated character of the two degree programs allows participants to acquire a wide range of knowledge and skills related to sustainability. What is more, the Master's course described is complemented with international experience, and the Doctoral course is complemented with training in and opportunities for practical reinforcement in the field.

The program started in 2005 as a two-year Master's course; its three-year doctoral course was created in 2007. As of October 2017, 36 students are enrolled in the Master's course and 36 students in the Doctoral course. The program has had students from more than 50 countries from all over the world. GPSS-GLI students also come from diverse academic backgrounds ranging from biology, civil engineering, economics, development studies, urban and rural planning, and numerous others. The program was established in the Graduate School of Frontier Sciences where new academic challenges are being developed through inter- or trans-disciplinary approaches, and collaborates very closely with the Integrated Research System for Sustainability Science (IR3S) based in The University of Tokyo. GPSS was reformed into GPSS-GLI in 2012 when the program was selected for Leading Graduate School Programs and started receiving a new funding from MEXT (the Japanese Ministry of Education, Culture, Sports, Science and Technology).

The type of education that GPSS-GLI offers is described as a T-shape education in which the horizontal line of "T" represents the breadth of knowledge on sustainability issues as well as practical skills for implementing projects, and the vertical line of "T" corresponds to the depth of knowledge specializing in one academic discipline obtained through a Master's thesis or Doctoral dissertation project. The program curriculum is designed in such way that the participants constantly revisit the T shape to avert becoming narrowly focused. The program believes this is necessary training for sustainability experts to become able to consider multiple dimensions of sustainability. Such training is done by courses on diverse topics in sustainability, by the advisory process by supervisors, and by the weekly GPSS-GLI seminar in which all participants have regular opportunities of mutual learning from the research progress presentations of others and of contributing to discussions in diverse disciplines.

More specifically, the GPSS-GLI program revolves around three key perspectives: wholistic, resilient, and trans-boundary. The holistic perspective implies a bird's-eye view that provides a combined view of an overarching perspective and in-depth understanding of the human-nature relationships. The resilient perspective employs flexibility in process governance that enables both long-term concerns (e.g. climate change) and short-term concerns (e.g. natural disasters) to be addressed concurrently. Lastly, trans-boundary perspective provides a comparative approach from a global scale to a local scale bringing diverse people together to jointly address sustainability issues.

The GPSS-GLI curriculum consists of three core components: (i) lecture courses focusing on theories and concepts, methodologies, and a wide range of topics related to sustainability; (ii) practicum courses aiming at developing interpersonal skills, systems thinking perspectives, and field survey methods; and (iii) a comprehensive research process beginning with identifying a research problem, developing research framework, implementation and data collection, all of which are compiled into a Master's thesis or Ph.D. dissertation.

Among the three main components of the GPSS-GLI curriculum, field-based training units in practicum courses are unique. Students travel to locations where actual sustainability issues are ongoing where they gain on-ground experience as well as practical skills in understanding the complex structure of the issue, identify possible leverage points, and design possible interventions. These field-based courses have covered mercury poisoning caused by rapid industrialization, rural sustainability in an aging society, natural disaster management in coastal areas, poverty and nutrition issues, and smart city development to name a few.

How to deliver contents that facilitate program participants' obtaining holistic, resilient, and trans-boundary perspectives has been the major challenge for GPSS-GLI since its establishment. The authors, however, have observed informational changes among the participants throughout the history the program, field-based training, and individual thesis research activities for more than 10 years. Especially, their worldviews are challenged in the program-wide weekly seminar that contributes to examining the application of framing of sustainability issues in student research projects.

1.3   What to Frame as Sustainability Challenges

Sustainability science is a problem-based or solution-oriented science (Clark and Dickson 2003; Kates et al. 2001), and this implies that a process to define what to frame is a challenge that exists in all sustainability research. Sustainability is "a fundamentally ethical concept raising questions regarding the value of nature, responsibilities to future generations, and social justice" (Norton 2005), yet a limited discussion has been held on these normative dimensions of sustainability in sustainability science research. Those challenges related to problem-based or solution-oriented dimensions such as climate change, biodiversity loss, and resource depletion have undoubtedly been framed as key sustainability issues because these problems will result in serious threats to human beings. However, a new set of challenges has been observed that are more related to human society, and the emergence of these challenges can be seen as a result of socioeconomic development such as rapid urbanization, mass production and consumption, and heavy transportation. Although these challenges are included in the UN's Sustainable Development Goals (SDGs), reviewing what we are framing as sustainability concerns in these challenges is critical to better understand what we are aiming to achieve through sustainable development.

The state of being sustainable tends to be seen as an absolute state of society that takes a balance between human system and nature systems (Giampietro 2002). When such a static perspective to sustainability is applied, a belief that lowering consumption of goods and services or reducing carbon emissions from our daily lives eventually leads our society to a sustainable state is commonly shared. Solution-oriented approaches tend to employ this static perspective and consider that sustainability can be achieved by designing systems in which agents follow the structured rules of the system. In reality, however, what sustainability means is to change gradually over time as people's value orientations change. Therefore, sustainability should not be seen as a fixed goal of our society, it is rather a process, and people only sustain what they frame as valuable based on their value propositions.

1.4   What Is Framing?

Framing is a common concept in many academic disciplines such as psychology, linguistics, sociology, communication and media studies, and political science. Framing explains how people perceive and interpret particular topics or events with the social norms, values, and assumptions that they apply in all situations (Benford and Snow 2000; Goffman 1974). When a majority of the general public applies one particular framing to one particular topic, then it provides explanations of why this topic matters, who is responsible, and what measures should be taken (Gamson et al. 1989; Price et al. 2005). Utilizing such characteristics of framing, reframing is sometimes used to set alternative perspectives to topics and events with particular meaning that the person or group would like to propose (Jarratt and Mahaffie 2009).

One main premise of framing theory is based on a constructivism perspective which realizes multiple ways of viewing and constructing the world (Chong and Druckman 2007). Hence, multiple framings by different groups of people in our society always exist, and how the concept of sustainability is framed is also multiple. To integrate such multiplicity of framing present among different actor groups, sustainability scientists facilitate collaborations through broad inter- and trans-disciplinary initiatives (Bammer 2005; Leach et al. 2010).

In sustainability science, framing is an important concept that examines the process to determine what is worthwhile to sustain in line with the direction of sustainable development. Sustainability fundamentally contains a normative dimension, and such framing is built upon social values and individual beliefs. Answering the core questions of sustainability—sustain what, for whom, how long, and at what cost—reflects our orientations in the framing process of particular topics. In order to move the discussion on framing in sustainability science forward, developing a conceptual framework that cautions us of key elements to consider is critical. To conclude the chapter, the authors propose their conceptual framework of key elements for visualizing transformation to a sustainable society.

1.5 Framework to Visualize Transformation to a Sustainable Society

Based on the experience of operating the GPSS-GLI program over the last ten years including the initial three years as GPSS, the authors have developed a conceptual framework that encompasses key elements that must be examined when discussing transformation towards sustainable society in a research or action plan. Fig. 1 summarizes those key elements for framing sustainability issues (shown as (1) *Framing complexities*) and suggests possible means (shown as (2) *Transformation channels*) to lead the current state of society to a sustainability pathway.

When addressing one sustainability issue, the authors argue that Holistic Treatment (top-down approaches) and Trans-boundary Thinking (bottom-up approaches) need to be applied jointly in order to (i) analyze the embedded complexity within the structure of the issue, (ii) develop action plans towards a sustainable society incorporating the uncertainty in this action planning process, and (iii) evaluate the entire framing process from issue identification, action plan development, and implementation. Holistic treatment and trans-boundary thinking are perspectives that support each other, an interaction which is necessary for examining one sustainability issue from multiple angles and for visualizing how proposed actions will unfold.

Holistic Treatment, which represents the upper half of the framework, is based on a systems perspective and depicts cause-and-effect relationships among the factors and agents related to an issue. In contrast, the Trans-boundary Thinking, which represents the bottom half of the framework, is based on individual-case perspectives that reflect the locality of a particular community or stakeholder group.

Such local perspective often does not match well with the global and national level sustainability goals. Therefore, the process to articulate local perspectives and reinterpret the global sustainability manifestations in individual cases is an important process when linking the Holistic Treatment and the Trans-boundary Thinking. Sustainability experts are expected to facilitate the communication among diverse stakeholder groups and supplement relevant information and knowledge to ensure the convergence of Holistic Treatment and Trans-boundary Thinking perspectives.

When a group of stakeholders addresses one sustainability issue, analyzing the complexity embedded in the issue is the first step. Complexity is understood as a system with parts, feedback, and non-linear and linear relationships (Ladyman et al. 2013). To frame the complexity of sustainability issues, several key factors must be examined. For example, the authors' framework suggests temporal and spatial dynamics of the issue (temporal diversity to long-term trends, local to global perspectives), legal, political, and institutional dimensions of the issue (intergenerational equity, institutional structure for concrete actions), world views and paradigms (differences in how to understand reality) [shown as (1) in Fig. 1]. Which framing becomes more helpful in analyzing the complexity depends on the nature of the discussed issue. Paying close attention to what kind of framing to be applied during issue identification, however, is critically important. This is because every framing process applies different assumptions, principles, and views to the issue; and it sets what topics are to be viewed as important, and what topics should be addressed or not. Those stakeholders who initiate actions aiming for a sustainable society must be able to see an issue from different angles by applying multiple framings.

When a group of stakeholders addresses one sustainability issue, analyzing the complexity embedded in the issue is the first step. Complexity is understood as a system with parts, feedback, and non-linear and linear relationships (Ladyman et al. 2013). To frame the complexity of sustainability issues, several key factors must be examined. For example, the authors' framework suggests temporal and spatial dynamics of the issue (temporal diversity to long-term trends, local to global perspectives), legal, political, and institutional dimensions of the issue (intergenerational equity, institutional structure for concrete actions), world views and paradigms (differences in how to understand reality) [shown as (1) in Fig. 1.1]. Which framing becomes more helpful in analyzing the complexity depends on the nature of the discussed issue. Paying close attention to what kind of framing to be applied during issue identification, however, is critically important. This is because every framing process applies different assumptions, principles, and views to the issue; and it sets what topics are to be viewed as important, and what topics should be addressed or not. Those stakeholders who initiate actions aiming for a sustainable society must be able to see an issue from different angles by applying multiple framings.

Once the structure of the issue is analyzed, possible actions for achieving a sustainable society are proposed. Yet, the authors argue that there are many steps in between the issue identification and action planning as shown in Fig. 1.1. The process of proposing actions must be done by combining a backcasting approach based on Holistic Treatment and process management base on Transboundary Thinking. A backcasting approach requires clear images of ideal goals or states, often they can

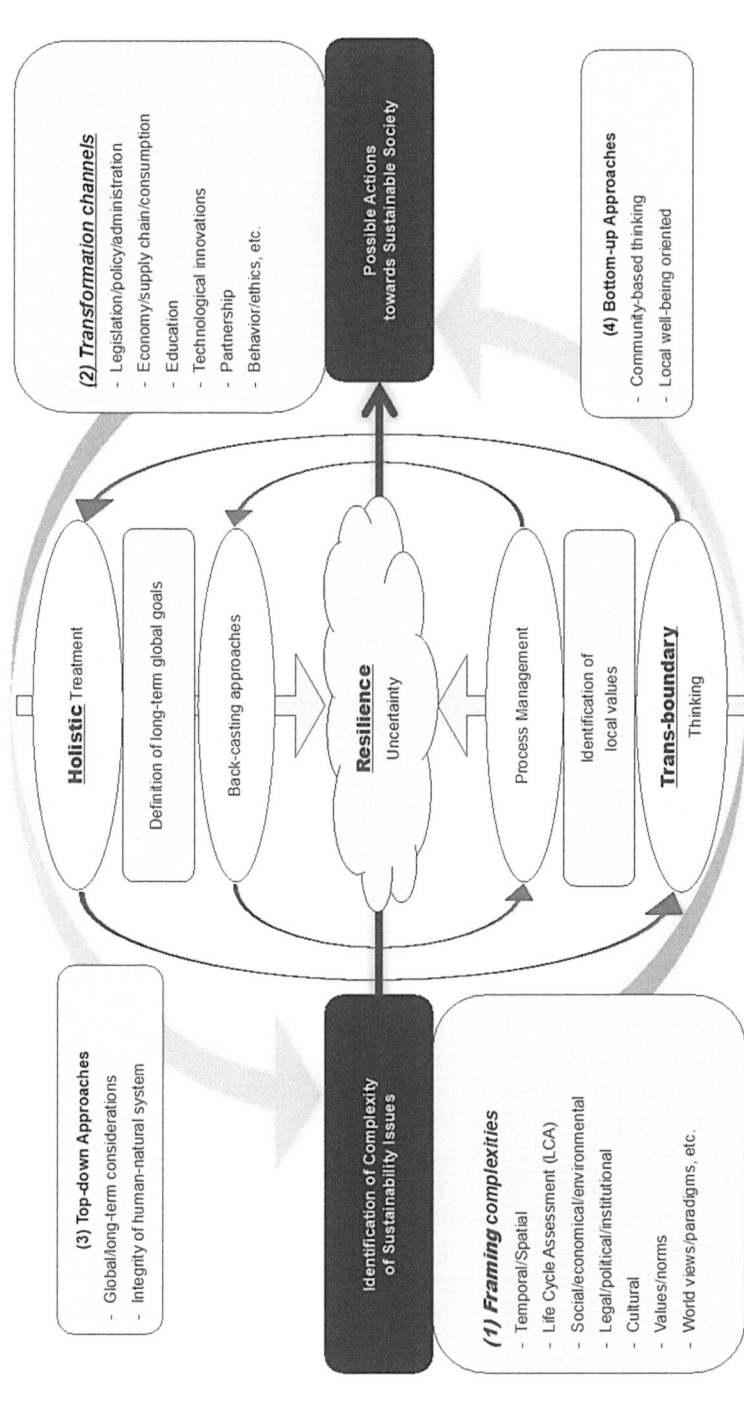

Fig. 1.1 A Conceptual framework of key elements for visualizing transformation to a sustainable society

Modified from Mino T et al. (2016) in "Sustainability Science: Field Methods and Exercises", Springer International Publishing Switzerland, Esteban M et al (Ed)

be linked with the global sustainability agenda such as the Sustainable Development Goals (SDGs). In contrast, process management aims at enhancing local values shared by fixed members, and these values can be identified by a Trans-boundary Thinking approach. The authors argue combining Holistic Treatment based on systems perspective and Trans-boundary Thinking based on the local perspective of each case is an essential methodology to incorporate the global sustainability agenda and the relevance in individual cases. This methodology enables researchers and stakeholders to develop a more resilient action plan that is ready to accommodate uncertainty.

Those proposed actions to achieve a sustainable society appear in various forms. Some of the Transformation channels are registration, policy and administration; economy, supply chain; and consumption, and technological innovations (shown as (2) in Fig. 1.1). These channels are further narrowed down to concrete actions to respond to identified challenges.

After actions are proposed through particular transformation channels, the framework suggests additional steps to examine how the proposed actions can be situated within factors in top-down approaches (global/long-term consideration, integrity of human-natural system) [shown as (3) in Fig. 1.1]. This process not only makes the linkage with global sustainability agenda (e.g. SDGs) explicit, but also verifies the transferability of the proposed actions. Following this verification step by holistic treatment perspective, the proposed actions must also be checked by the factors in the bottom-up approaches (community-based thinking, local well-being orientation) [shown as (4) in Fig. 1.1]. These two steps functions simultaneously and serve as an evaluation part for the entire process and its possible influence at various scales.

In summary, the framework serves as a guideline for researchers and stakeholders to (i) analyze the complexity of sustainability issues through multiple framings, (ii) apply holistic treatment and trans-boundary thinking in the process of developing action plans, and (iii) evaluate the proposed actions from the perspectives of both top-down approaches and bottom-up approaches. In many of the actual cases, the sustainability issues are already analyzed and action plans are being implemented by the time a theoretical framework is applied. Therefore, the proposed framework is to be introduced into the process at any time. For example, the framework can be used to evaluate the outputs of conducted actions first, then further utilized to re-examine the applied framing to understand the complexity to the addressed issue before the second round of concrete actions is taken. The authors believe this framework incorporates key elements of framing in sustainability science thus far. This framework, however, must still incorporate the concept of resilience in its goal-setting process and in the uncertainties in the process management, which have not been discussed in detail in this paper. Sustainability experts must be trained with knowledge and skills to perform the suggested steps when utilizing this framework in sustainability research and action projects.

1.6 Scope and Structure of this Book

This book aims at examining different types of framing applied by scholars to sustainability research. In so doing, this book provides an overall picture of sustainability research by scholars from different academic backgrounds (i.e., representing different ontologies and epistemologies). As efforts continue in achieving sustainable development goals and trying to guide society to sustainability, realizing different intentions behind each framing and being open to negotiation as well as cooperation are important for sustainability experts. The first step in starting such an approach is gaining understanding of one another's framings.

References

Bammer G (2005) Integration and implementation sciences: building a new specialization. Ecol Soc 10(2). Retrieved from http://www.ecologyandsociety.org/vol10/iss2/art6/

Benford RD, Snow DA (2000) Framing processes and social movements: an overview and assessment. Annu Rev Sociol 26:611–639. Retrieved from http://www.jstor.org/stable/223459

Bettencourt LMA, Kaur J (2011) Evolution and structure of sustainability science. P Nat Acad Sci USA 108(49):19540–19545. https://doi.org/10.1073/pnas.1102712108

Chapin FS, Pickett ST, Power ME, Jackson RB, Carter DM, Duke C (2011) Earth stewardship: a strategy for social–ecological transformation to reverse planetary degradation. J Environ Stud Sci 1(1):44–53. https://doi.org/10.1007/s13412-011-0010-7

UNESCO Chetto AM (ed) (2000) World conference on science: Science for the twenty-first century: A New Commitment. Paris Retrieved from https://www.google.co.jp/url?sa=t&rct=j-&q=&esrc=s&source=web&cd=1&cad=rja&uact=8&ved=0ahUKEwjBkdXdienRAhUExrw KHW_aB-EQFggcMAA&url=http%3A%2F%2Funesdoc.unesco.org%2Fimages%2F0012% 2F001207%2F120706e.pdf&usg=AFQjCNEf9N611stsq2oJUzorEQGCgLMBXA

Chong D, Druckman JN (2007) Framing theory. Annu Rev Polit Sci 10:103–126. https://doi. org/10.1146/annurev.polisci.10.072805.103054

Clark WC (2007) Sustainability science: a room of its own. P Natl Acad Sci USA 104(6):1737. https://doi.org/10.1073/pnas.0611291104

Clark WC, Dickson NM (2003) Sustainability science: the emerging research program. P Natl Acad Sci USA 100(14):8059–8061. https://doi.org/10.1073/pnas.1231333100

Fiksel J (2003) Designing resilient, sustainable systems. Environ Sci Technol 37(23):5330–5339. Retrieved from http://www.ncbi.nlm.nih.gov/pubmed/14700317

Gamson WA, Mondigliani A, Modigliani A (1989) Media discourse and public opinion on nuclear power: a constructionist approach. Amer J Sociol 95(1):1–37. https://doi.org/10.1086/229213

Giampietro M (2002) Complexity and scales: the challenge for integrated assessment. Integrat Asses 3:247–265. Retrieved from http://journals.sfu.ca/int_assess/index.php/iaj/article/ viewArticle/33

Goffman E (1974) Frame analysis: an essay on the organization of experience. Northeastern University Press, York, Pennsylvania

Hiramatsu M (2012) Sasuteinabiriti-gaku ni okeru shakaigakuteki shiten (Dai-ippou) [Sociological Perspective in Sustainability Science I: From IR3S (サステイナビリティ学における社会学的視点(第1報))]. 名古屋女子大学 紀要, 58, 101–108. Retrieved from https://www. google.co.jp/url?sa=t&rct=j&q=&esrc=s&source=web&cd=1&cad=rja&uact=8&ved=0a hUKEwit6K3i7uvRAhWHTbwKHVOvBdMQFggcMAA&url=https%3A%2F%2Fnago ya-wu.repo.nii.ac.jp%2F%3Faction%3Drepository_action_common_download%26item_ id%3D1313%26item_no%3D1%26attri

Jarratt J, Mahaffie JB (2009) Reframing the future. J Fut Stud 13(4):5–12. Retrieved from http://jfs-digital.org/articles-and-essays/2009-2/vol-13-no-4-may/articles-essays/reframing-the-future/

Jerneck A, Olsson L (2011) Breaking out of sustainability impasses: how to apply frame analysis, reframing and transition theory to global health challenges. Environ Innov Societ Trans 1(2):255–271. https://doi.org/10.1016/j.eist.2011.10.005

Jerneck A, Olsson L, Ness B, Anderberg S, Baier M, Clark E, Hickler T, Hornborg A, Kronsell A, Lövbrand E, Persson J (2011) Structuring sustainability science. Sustain Sci 6(1):69–82. https://doi.org/10.1007/s11625-010-0117-x

Kajikawa Y (2008) Research core and framework of sustainability science. Sustain Sci 3:215–239. https://doi.org/10.1007/s11625-008-0053-1

Kajikawa Y, Tacoa F, Yamaguchi K (2014) Sustainability science: the changing landscape of sustainability research. Sustain Sci 9(4):431–438. https://doi.org/10.1007/s11625-014-0244-x

Kates RW (2001) Environment and development: sustainability science. Science 292(5517):641–642. https://doi.org/10.1126/science.1059386

Kates RW (2011) What kind of a science is sustainability science? P Natl Acad Sci USA 108(49):19449–19450. https://doi.org/10.1073/pnas.1116097108

Kates RW, Clark WC, Corell R, Hall JM, Jaeger CC, Lowe I, McCarthy JJ, Schellnhuber HJ, Bolin B, Dickson NM, Faucheux S, Gallopin GC, Grübler A, Huntley B, Jäger J, Jodha NS, Kasperson RE, Mabogunje A, Matson P, Mooney H, Moore III B, O'Riordan T, Svedin U (2001) Sustainability Science Science, 292(5517):641–646. https://doi.org/10.1126/science.1059386

Kinzig A, Ryan P, Etienne M, Allison H, Elmqvist T, Walker BH (2006) Resilience and regime shifts: Assessing cascading effects. Ecol Soc 11(1) Art 20. Retrieved from http://www.ecologyandsociety.org/vol11/iss1/art20/

Komiyama H, Takeuchi K (2006) Sustainability science: building a new discipline. Sustain Sci 1(1):1–6. https://doi.org/10.1007/s11625-006-0007-4

Ladyman J, Lambert J, Wiesner L (2013) What is a complex system? Eur J Philos Sci 3:33):33–33):67. https://doi.org/10.1007/s13194-012-0056-8

Lang DJ, Wiek A, Bergmann M, Stauffacher M, Martens P, Moll P, Swilling M, Thomas CJ (2012) Transdisciplinary research in sustainability science: practice, principles, and challenges. Sustain Sci 7(S1):25–43. https://doi.org/10.1007/s11625-011-0149-x

Leach M, Scoones I, Stirling A (2010) Dynamic sustainabilities: technology, environment, social justice. Routledge, New York

van der Leeuw S, Wiek A, Harlow J, Buizer J (2012) How much time do we have ? Urgency and rhetoric in sustainability science. Sustain Sci 7:115–120. https://doi.org/10.1007/s11625-011-0153-1

Lubchenco J (1998) Entering the century of the environment: a new social contract for science. Science 279(23):491–497. https://doi.org/10.1126/science.279.5350.491

Martens P (2006) Sustainability: science or fiction? Sustain Sci Pract Policy 2:(1):36–41. https://doi.org/10.1080/15487733.2006.11907976

Morse S (2010) Sustainability: a biological perspective. Cambridge University Press, London/New York

Norton B (2005) Sustainability: a philosophy of adaptive ecosystem management. University of Chicago Press, Chicago. Retrieved from http://press.uchicago.edu/ucp/books/book/chicago/S/bo3641681.html

Olsson L, Jerneck A, Thoren H, Persson J, O'Byrne D (2015) Why resilience is unappealing to social science: theoretical and empirical investigations of the scientific use of resilience. Sci Advan 1(4):e1400217. https://doi.org/10.1126/sciadv.1400217

Ostrom E, Janssen MA, Anderies JM (2007) Going beyond panaceas. P Natl Acad Sci USA 104(39):15176–15178. https://doi.org/10.1073/pnas.0701886104

Phillips J (2010) The advancement of a mathematical model of sustainable development. Sustain Sci 5(1):127–142. https://doi.org/10.1007/s11625-009-0103-3

Price V, Nir L, Cappella JN (2005) Framing public discussion of gay civil unions. Publ Opin Qtly 69(2):179–212. https://doi.org/10.1093/poq/nfi014

Rosen R (2005) Life itself: a comprehensive inquiry into the nature, origin, and fabrication of life. Columbia University Press, New York

Satanarachchi N, Mino T (2014) A framework to observe and evaluate the sustainability of human-natural systems in a complex dynamic context. Springerplus 3(618). https://doi.org/10.1186/2193-1801-3-618

Schoolman ED, Guest JS, Bush KF, Bell AR (2012) How interdisciplinary is sustainability research? Analyzing the structure of an emerging scientific field. Sustain Sci 7(1):67–80. https://doi.org/10.1007/s11625-011-0139-z

Shahadu H (2016) Towards an umbrella science of sustainability. Sustain Sci 11(5):777–788. https://doi.org/10.1007/s11625-016-0375-3

Spangenberg JH (2011) Sustainability science: a review, an analysis and some empirical lessons. Environ Conserv 38(3):275–287. https://doi.org/10.1017/S0376892911000270

Sterman JD (2012) Sustaining sustainability: creating a systems science in a fragmented academy and polarized world. In: Weinstein MP, Turner RE (eds) Sustainability science: the emerging paradigm and the urban environment. Springer-Verlag New York, New York, pp 21–58. https://doi.org/10.1007/978-1-4614-3188-6

Swart RJ, Raskin P, Robinson J (2004) The problem of the future: sustainability science and scenario analysis. Global Environ Chang 14(2):137–146. https://doi.org/10.1016/j.gloenvcha.2003.10.002

de Vries BJM (2013) Sustainability science. Cambridge University Press, Cambridge, UK. Retrieved from http://www.cambridge.org/jp/academic/subjects/earth-and-environmental-science/environmental-science/sustainability-science?format=PB&isbn=9780521184700

WCED (1987) Our common future. Oxford University Press, Oxford

Weinstein MO, Turner RE (eds) (2010) Sustainability science: the emerging paradigm and the urban environment. Sustain: Sci Pract Pol 6(1):1–5. Retrieved from files/2218/Weinstein and Turner – 2012 – Sustainability Science The Emerging Paradigm and.pdf

Wiek A, Lang DJ (2016) Transformational sustainability research methodology. In: Heinrichs H, Martens P, Michelsen G, Wiek A (eds) Sustainability science an introduction, 1st edn. Springer, Dordrecht/Heidelberg/New York/London. https://doi.org/10.1007/978-94-017-7242-6

Wiek A, Ness B, Schweizer-Ries P, Brand FS, Farioli F (2012) From complex systems analysis to transformational change: a comparative appraisal of sustainability science projects. Sustain Sci 7(Supplement 1):5–24. https://doi.org/10.1007/s11625-011-0148-y

Yarime M (2013) Exploring sustainability science: knowledge, institutions, and innovation. In: Sustainability science: a multidisciplinary approach. United Nations University Press, Tokyo, pp 98–111. https://doi.org/10.18356/858f44f9-en

Chapter 2
Theoretical and Methodological Pluralism in Sustainability Science

Anne Jerneck and Lennart Olsson

Abstract Sustainability science is an integrative scientific field embracing not only complementary but also contradictory approaches and perspectives for dealing with an array of sustainability challenges.

In this chapter we distinguish between pluralism and unification as two main and distinctly different approaches to knowledge integration in sustainability science. To avoid environmental determinism, functionalism, or overly firm reliance on rational choice theory, we have reason to promote pluralism as a way to better tackle sustainability challenges. In particular we emphasise two main benefits of taking a pluralist approach in research: it opens up for collaboration, and ensures a more theoretically informed understanding of society.

After a brief introduction to how we interpret the field of sustainability science, we discuss ontology, epistemology and ways of understanding society based on social science theory. We make three contributions. First, we identify important reasons for the incommensurability between the social and natural sciences, and propose remedies for how to overcome some of the difficulties in integrative research. Second, by suggesting a frame that we call 'social fields and natural systems' we show how sustainability science will benefit from drawing more profoundly on – and thus more adequately incorporate – a social science understanding of society. As such, the frame is a foundation for pluralism. Third, by suggesting a new theoretical typology, we show how sustainability visions and pathways are associated with particular theoretical and methodological perspectives in geography, political science, and sociology; and how that matters for research and politics addressing sustainability challenges. The typology can be used as a thinking tool to frame and reframe research.

Keywords Incommensurability · Knowledge integration · Social fields · Social change · Unification

A. Jerneck · L. Olsson (✉)
Lund University, Lund, Sweden
e-mail: lennart.olsson@lucsus.lu.se

2.1   Introduction – What Has Become of Sustainability Science?

The fact that sustainability science is 'dealing with interconnected problems' (Kauffman and Arico 2014) requires that researchers in the field take a comprehensive, integrated, and participatory approach to science and reality (Sala et al. 2013). In line with this explicit ambition to integrate knowledge – across scales, sectors, and substance domains; and across the divides of nature-society, science-society and knowledge-action – sustainability science must build on several foundational disciplines and inherently advocate theoretical and methodological pluralism (Persson et al. 2018a, b).

The focus in early sustainability science was threefold. It centred on elucidating nature–society interaction, providing scientific knowledge for sustainability, and elaborating normative discussions on sustainability. For this purpose it gave prominence to problem-driven and solutions-oriented research on human-environmental interaction – or what some call socio-ecological dynamics – while envisioning sustainable futures. In that endeavour, Cash et al. (2003) asked sustainability scientists to apply *credibility, legitimacy*, and *saliency* in research, especially when it comes to *data and methods, focus and findings, and outreach and solutions*. These quality features are not necessarily exclusive to sustainability science but remind us of critical theory which also aims at social change and on which ideas sustainability science can build. What is more typical is perhaps that sustainability scientists are expected to ask pertinent questions: what to sustain, for whom, for how long, and at what benefit or cost?

In the absence of any universal criteria that define sustainability science (Shahadu 2016, p 2) we wish to point out some common denominators that unite the field. Starting from interdisciplinarity while striving for transdisciplinarity, sustainability science takes a broad approach to understanding and improving social life *within* the broader context of earth's life support systems. Sustainability science researchers are expected to pose integrated questions that capture human-environment conditions; and while doing so, develop theoretical and methodological frames for overcoming constraining differences in research methods and procedures across disciplinary boundaries. A sustainability science community would, ideally, bring together researchers with a variety of disciplinary (or interdisciplinary) repertoires to discuss and negotiate the multiple meanings of concepts and phenomena that are significant in sustainability science research – and crucial for sustainability. Beyond that, and again ideally, stakeholders other than academics would be called upon at various stages in the research process to inspire problem formulations and help sharpen the focus on conditions for and implications of human-environment interaction and interdependence.

At the risk of denying, ignoring, or limiting diversity and pluralism, some sustainability scientists have called for a certain degree of standardisation in sustainability science. This **can** take the form of a 'core set' of assumptions, concepts, ideas, and understandings that would speak across research studies (Shahadu 2016).

This was attempted in the early days of sustainability science when pioneering front-figures launched the research field through a suite of urgent questions and suggested actions (Kates 2011; Kates et al. 2001). These intentions and questions have since then been followed up by articles on framing, knowledge structuring, and the many methodological concerns in sustainability science research (Jerneck and Olsson 2011; Jerneck et al. 2011; Spangenberg 2011; Thoren 2015; Wiek et al. 2012). In that mission scholars have emphasised the need for acknowledging values and social learning processes when imagining desirable futures (Miller et al. 2013). In this chapter, we take the knowledge structuring further by calling for ontological, epistemological and theoretical awareness in problem formulation; and by providing a thinking tool to compare, frame, and juxtapose theoretical approaches in sustainability science.

To further expand and refine the field, researchers in sustainability science must continually discuss the significance of its substance, scrutinise its approaches, and confront internal conflicts while searching for synergies (Isgren et al. 2017). As suggested by Shahadu (2016), we can do this under the heading of an 'umbrella science' that is distinct in focus while inclusive in welcoming subfields (see also (Miller et al. 2013). For that we could consider sustainability science research in terms of its mission and mandate, its achievements, and its challenges and conflicts (Isgren et al. 2017). Further, Lang et al. (2012) call for evaluative, qualitative and quantitative meta-studies of sustainability to make use of existing evidence and experience more systematically.

In the actual practice of doing sustainability science, we suggest with many others that researchers in the field should pay particular attention to three core aspects – collaboration and communication, reflexivity, and research design – and below we offer some justifications.

2.1.1 Collaboration and Communication

Integrated research that seeks to build knowledge across divides and between disciplinary domains can be 'a response to the complex demands of the modern world' as well as 'a source of competitive advantage' (Siedlok and Hibbert 2014). Notably, integrated collaboration between communities enhances the understanding of 'what is the problem' while also advancing learning and innovation around 'what can and should be done', 'within what time frame', and 'by whom'. Bearing that process in mind, studies on integrated research show that a high degree of communication and interaction is necessary in creating such diverse groups (Hage and Hollingsworth 2000).

2.1.2   Reflexivity

Sustainability science is defined less by its disciplinary content – and more by its purpose, the problems it studies, the types of solutions it seeks, its applicability (Clark 2007), and the role of reflexivity in interdisciplinary and transdisciplinary research (Spangenberg 2011). As a core competence in integrative research, reflexivity means to question assumptions such as those about the ability to predict future events, the objectivity of the observer, and the value neutrality of science (Spangenberg 2011). It will also require the acceptance of ignorance, uncertainty, and the impossibility of knowing all relevant aspects of evolving systems or foreseeing emergent system properties (Spangenberg 2011, p 279). Finally, and owing to its attributes, sustainability science is 'a shared learning endeavor' within which participants must **also** include the learning from failures and setbacks (Barth and Michelsen 2013).

2.1.3   Research Designs

An appropriate research design for sustainability science must be **general** enough in scope to include various sustainability challenges and contexts, **flexible** enough to include a process of ongoing revision that allows (or even ensures) a reconstruction of methods and practices when needed (van Kerkhoff 2014, p 149), and **specific** enough 'to offer genuine guidance' (van Kerkhoff 2014, p 145). In the process of developing a (format for a good) research design, inspiration can be gained from various sources. Onto-epistemological reflections are inspirational for identifying and defining relevant and interesting research problems and for selecting particular units of analysis to focus on. Theoretical-conceptual reflections are helpful for developing concepts, theories, and perspectives that may promote our understanding of a broader range of issues (Salas-Zapata et al. 2013). This can also include thoughts on how to plan and organise stakeholder participation, how to deal with uncertainty and the limits to knowledge and data construction, and what to expect from explorative science. *Instrumental-methodological* reflections will help us *apply* theories and concepts to real-world conditions, events, and situations – and thus facilitate the analysis.

2.2   Ontology – On Reality, Systems and Fields

Ontology is concerned with assumptions, claims, and questions about what exists in the world, how reality presents itself, and to what extent that reality is observable. Differences in ontology and epistemology constitute a main obstacle to the integration of knowledge across the boundaries of scientific disciplines (Jerneck et al.

2011). By knowledge integration we mean a process where the best available knowledge from two or more scientific disciplines or fields is used to understand a complex problem. A central challenge to knowledge integration in sustainability science is how to deal with seemingly incompatible assumptions deriving from varied ontological claims in the natural and social sciences. This involves a concern for how to ensure that the best available social science knowledge is used in combination with the best available knowledge in natural sciences, engineering, and medicine. A concern following from that is how to study issues such as the consequences of climate change impacts on society without resorting to environmental determinism. To clarify here, environmental determinism is a foundational element of colonialism referring to the belief that natural conditions shape societies. Another concern refers to the frequent use of indicators which illustrates a tendency to emphasize that reality is observable and measurable.

Systems and system boundaries are core ontological components of the natural sciences, both in theory such as stocks and flows models and in practice for describing a quantifying bio-geo-chemical fluxes. Meanwhile, in contemporary social science inherent obstacles to systems thinking abound. Researchers studying social phenomena based on social theory may be reluctant to use systems as an ontological description of society but may decide to use it analytically to study a specific aspect of the economy, polity, or society such as the tax system, the party system, the energy system, or the social security system. However, the neo-liberal zeitgeist has made it so natural to speak in systems terms such as resilience and self-organization of socio-ecological systems as well as adaptive management because of climate change that we fail to see the contradicting political forces behind it. But following Colin Hay (2002, p 3) 'All events, processes and practices which occur within the social sphere have the potential to be political and, hence, to be amenable to political analysis'. What makes an analysis political is its focus and emphasis on 'the political aspect of social relations' and in particular the 'attention to the power relations implicated in social relations' (Hay 2002, p 3). This implies, for example, that the 'sociology of structural inequality' is a subject of political analysis (Hay 2002, p 3). If translated into a sustainability science context it means that the many socially, spatially and temporally uneven impacts of and responses to climate change ought to be studied while remembering that politics are concerned with 'the distribution, exercise and consequences of power' (Hay 2002, p 3).

To bridge ontological barriers in sustainability science research while avoiding not only the risk of scientific imperialism associated with unification meaning that one discipline dominates another but also the risk of de-politicisation of socio-ecological issues, we suggest the use of two explicit ontological assumptions: *social fields and natural systems* (Olsson and Jerneck 2018). Below, we will return to a more specific description of social fields and a discussion of how to justify its use. Such an approach has the potential to overcome ontological differences between the social and natural sciences, and is also useful for avoiding three common weaknesses of knowledge integration across the natural and social sciences as mentioned above, namely the use of environmental determinism, functionalism, and rational choice theory to explain social change. In combination, they often result in a de-politicisation of environmental problems and even scientific justification of particular policies (Olsson et al. 2015; Wellstead et al. 2016; Newton 2016). Jared Diamond's stories about human development and collapse (Diamond 1999, 2005) and Jeffrey Sachs' explanation of underdevelopment (Sachs and Warner 1995) exemplify a resurgence of determinism, or neo-environmental determinism (Sluyter 2003).

2.3 Epistemology – On Pluralism and Unification in Sustainability Science

Epistemology is concerned with assumptions, claims, and questions about how to gain knowledge about the world, who is a 'knower', and how to combine or integrate different types of knowledge. We identify two types of scientific knowledge integration – pluralism and unification (Olsson et al. 2015; Geels et al. 2016). *Scientific pluralism* is a process in which several disciplines contribute particular theories, methods, and/or questions to address or solve a problem. According to scientific pluralism, the ultimate goal of scientific inquiry is not (necessarily) to establish a single theory (Kellert et al. 2006). Rather, pluralism is useful in situations where no unified theories are available to explain a phenomenon or where the phenomenon can only be explained through the lens of multiple theories (Dupre 1991; Mitchell 2009). Undoubtedly this is the situation in a comprehensive context such as that of climate change or geopolitics.

In contrast, *unification* may easily result in scientific imperialism, a process usually thought of as an illicit infringement such as when one discipline attempts to explain phenomena or solve problems in a domain belonging to or associated with another discipline (Dupré 1994, 2001; Mäki 2013). Serious cases of scientific imperialism are reductive in the sense that they aim, or tend, to exclude alternative (even compatible) explanations and solutions (Clarke and Walsh 2009; Midgley 1984; Thoren 2015) resulting in a situation where inferior explanations or problem solutions outcompete superior ones (Thoren 2015). All kinds of unification are not necessarily imperialist (in this negative sense), but there is always reason to worry about imperialism in situations where a single theory (or discipline) is claimed to account for major or persistent social problems such as inequality, poverty, and social unrest, or for complex phenomena such as geopolitics or climate change impacts on society and its responses to that. In contrast, in the context of geopolitics, pluralism has not only scientific value, but can also help sustain cultural, ecological, and social diversity (Norgaard 1989). In practice, this can be pursued through research that harnesses both scientific and indigenous knowledge while also seeking to reconcile them (Agrawal 1995; Parsons et al. 2017; Persson et al. 2018b).

2.4 Ways of Understanding Society

2.4.1 Theory

Theory serves as a main guide to empirical exploration. It serves to simplify reality and to describe and explain it in terms that are appropriate for – and thus comparable between – different contexts. Theory can be descriptive, prescriptive, or predictive and can be used to challenge stated and unstated assumptions. Theories are characterised by their distinct perspectives and are (often) conceived of and expressed to represent a special subject-position or vantage-point (Hay 2002, p 24). This means that theory is not necessarily neutral, but often imbued by values and interests.

Inspired by Colin Hay (2002) we will discuss three particular issues relating to theory: the role of consensus and conflict theory in sustainability science; the tension between parsimony (the world is assumed to be simple and can be abstracted, explained, and predicted) and complexity (the world is assumed to be nuanced and can mainly be described concretely and only with some degree of plausibility); and finally, the interaction between agency and structure in society.

2.4.2 Consensus or Conflict

An important source of incommensurability between the social sciences and most natural sciences interested in the processes of environmental degradation, exploitation, or pollution, is how society is understood. We can identify two main types of approaches to understand society, resembling what in sociology is called consensus theory versus conflict theory.

According to consensus theory, shared norms and values are the foundation of a stable harmonious society in which social change is slow and orderly. For example, when using the concept of coupled social ecological systems, resilience can be seen as the equivalent of stability, harmony, and the 'good norm' (Olsson et al. 2015; Hatt 2013). In contrast, conflict theories emphasise competing interests between groups in society meaning that social order is maintained by (material or discursive) manipulation and control by dominant and powerful groups, and that transformational change develop from the tensions between these groups and the redistribution of power (Meadowcroft 2011). According to conflict theory, institutions are shaped by existing power imbalances, values, and social stratifications in society. This implies that governance is executed and understood differently in consensus theories versus conflict theories.

2.4.3 Parsimony or Complexity

The choice between complexity and parsimony is important in the selection of analytical perspective (Hay 2002, p 29). A parsimonious model is as simple as possible but explains as much as possible. However, at some point the merits of parsimony may be outweighed by greater complexity (Hay 2002, p 32). At one end of a spectrum pure description may capture real complexity without explaining much; whereas at the other end, abstract theoretical reasoning may be forceful in explaining and predicting a lot without capturing layers, nuances and crucial details (Hay 2002, p 35).

Seeking to preserve complexity while capturing specificity, constructivist, and institutionalists proceed with theory in close dialogue with data and details to piece together theoretically informed and empirically grounded historical narratives (Hay 2002, p 47). They suggest or establish the pre-conditions, conditions, and mechanisms of change by studying the interplay between ideas, institutions including their values, and interests pursued by actors. In so doing, they are inclined to acknowledge complexity, identify sequencing, and consider timing – all of which **are** enabled by methods of process-tracing, process-elucidation and a general open-ended approach to processes (Hay 2002, p 11). In sustainability science, constructivists and institutionalists are prone to locate and analyse the political aspects of the environment by considering how to value, prioritise, and sequence different social goals and sustainability pathways.

Acquiring and interpreting data implies a series of theoretical and methodological choices. Rather than taking regularity as a given and a basis for prediction, a constructivist or institutionalist would explore the conditions for and existence of both regularities and irregularities (Hay 2002, p 48). In such research considering whether conflict or cooperation is the norm in society is obviously important. And again, proponents of using indicators may have a tendency to seek readily observable data while also seeking regularity and stability in society, whereas those who emphasise the role of values may assume that society is divided by conflict and interests – and thus seek other types of data.

2.5 Ways of Understanding Agency, Behaviour, and Interaction

One important dividing line in the social sciences is how to define, explain, and understand human agency and behaviour, i.e. how people act and perform, the scope and limits of our agency, and based on what reasons people make decisions. As a starting point, structuralism tends to reduce social outcome to the workings of institutions and structures beyond the control of actors and their agency, whereas actor-oriented theory such as intentionalism (Hay 2002, p 55, Dessler 1989) tends to account for observable effects in purely agential terms.

In rational choice theory as the foundation of neoclassical economic theory, individuals make decisions based on maximising their own utility. The assumption of rational choice provides a reductionist basis for modelling the economy as a self-organising system. It also provides a scientific justification for the current proliferation of market-based instruments for ecosystem management. This is epitomised by The Economics of Ecosystems and Biodiversity initiative (TEEB) aiming to 'help decision-makers recognise, demonstrate and capture the values of ecosystem services & biodiversity' (Kumar 2010; Brown 2014).

Rational choice theory is widely used but contested in the social sciences. Other and more elaborated theories for explaining social behaviour have been formulated in sociology, such as various institutional theories and symbolic interactionism. In institutional theory, different scholars stress different aspects of social and economic interaction and relations (Mahoney 2000; Mahoney and Thelen 2010; Taylor 2011). In sharp contrast to rational choice theory, symbolic interactionism stresses social relations, contextual conditions, and subjective interpretation (Blumer 1986).

2.6   An Integrative Framework – Social Fields and Natural Systems

Inspired by American and French sociologists Neil Fligstein, Paul DiMaggio, Pierre Bourdieu and Loïc Wacquant, we suggest a new analytical framework for integrating knowledge across the natural and social sciences. As for now, we call it **Social Fields and Natural Systems** thus juxtaposing two ontological assumptions: the natural environment can be described in terms of systems, the social sphere is better described in terms of social fields (Olsson and Jerneck 2018). We argue that the approach has the potential to help researchers overcome ontological barriers between the social and natural sciences, and is particularly useful for avoiding the three common weaknesses in knowledge integration across the natural and social sciences that we mentioned earlier: the use of environmental determinism, functionalism, and rational choice as three theories attempting to explain social change.

In Earth System Science, the fundamental ontological assumption is that the world is a system. As long as the system is understood in primarily natural science terms such as an ecosystem, this is usually uncontroversial. Some ecologists claim that 'ecological and social domains of social-ecological systems can be addressed in a common conceptual, theoretical, and modelling framework' (Walker et al. 2006). This is the situation where a system ontology may come into conflict with ontological assumptions in the social sciences.

To Bourdieu, a *field* is a network of relations among actors and objects and their objective positions in the field (Ritzer 2011). John Levi Martin is another contemporary scholar who theorises fields. This quote describes his view (Martin 2003):

> I make the case that field theory is something quite different that has the potential to yield general but nontrivial insights into questions rightly deemed theoretical and to organize research in a productive fashion. Finally, field theory allows for the rigorous reflexivity that is necessary in all cases in which sociology attempts large-scale political and institutional analyses.

In their study of transnational migration, Levitt and Schiller (Levitt and Schiller 2004) applied field theory to highlight and study hidden institutions and social processes, and, importantly, challenge a routine notion of geographical scales (Levitt and Schiller 2004):

> The concept of social field also calls into question neat divisions of connection into local, national, transnational, and global. In one sense, all are local in that near and distant connections penetrate the daily lives of individuals lived within a locale.

Sociologists Fligstein and McAdam (2012) aim to construct a comprehensive and general theory of fields. Even if fields may lend similarities from systems and from institutional logics (Scott 1995), fundamental differences exist. They see *strategic action fields* as meso-level social orders which are the basic structural building blocks of modern political and organisational life. The identification and understanding of these *strategic action fields* are the basis for studying stability and change in society. Importantly, relations exist independently of whether people are aware of them or not, and whether people want them or not. Bourdieu was primarily interested in fields such as culture, education, and religion. In their general theory of fields, Fligstein and McAdam expand the notion of fields to become a more or less universal concept for studying social change and social order. In doing so, they expand the conceptual vocabulary and the horizon for what to study as a field.

The concept of incumbents and challengers was first introduced in field theory in the 1970s by William Gamson (1975) in his investigations of social movements. Incumbents have disproportionate power in or over a field and where the field in turn supports them. In contrast, challengers are less privileged in the field and are either in opposition to, or are more often suppressed by the field.

As an important argument against rational choice explanations of social change in a field, Fligstein and McAdam strongly argue that '*the material and the existential cannot be disentangled*' (Fligstein and Fligstein and McAdam 2012, p 49). They also stress the importance of social skills, defined as capacity for intersubjective thought and action in social relations. The concept of social skill is rooted in symbolic interactionism which rests on three main assumptions (Blumer 1986):

1. individuals act toward things on the basis of the meanings they ascribe to those things (i.e. things have no universal value in themselves),
2. the meaning of such things is derived from, or arises out of, the social interaction that one has with others and with society implying that decisions are primarily relational rather than individual, and
3. these meanings are handled in, and modified or recreated through, an interpretative process used by the individual in dealing with the things s/he encounters.

To exemplify the use of social fields and natural systems, we look at the issue of adaptation to current and future impacts of climate change. The number and severity of climatic extremes have clearly increased in recent years as a result of climate change (Field et al. 2012). Many of these events are associated with great losses of people, livelihoods, and property (Olsson et al. 2014) as well as with displacement and migration (Ionesco et al. 2016). The social responses to climate impacts are diverse and complex and do not follow any simple cause-effect pattern. Adaptation studies thus provide a good illustration of how multiple ontologies, i.e. fields and systems can support and promote the production of actionable knowledge.

To take an example, the climatic event affects a clearly defined geographical space or system. In the case of a flood, the impact is usually defined by the watersheds affected, i.e. a hydrological system. Using hydrological process models (e.g. SWAT, MIKE_SHE, or TOPMODEL (Devia et al. 2015)) the extent and severity of flood impacts can be understood and predicted. But neither the social repercussions nor all social drivers follow the natural system boundaries. Here strategic action fields can effectively be used to analyse and explain how social dynamics interact with the natural systems. In the figure below we try to illustrate how increasing frequency and intensity of floods, as a consequence of climate change in combination with social drivers, can be analysed through interacting multiple ontologies: systems for the natural science aspects (here represented by the Indus river basin in a systems model) and strategic action fields for the social aspects (here represented by climate politics in interacting horizontal and vertical fields). Some fields are interrelated and/or interact directly with some systems components, whereas others are independent, indirect, or diffuse (Fig. 2.1).

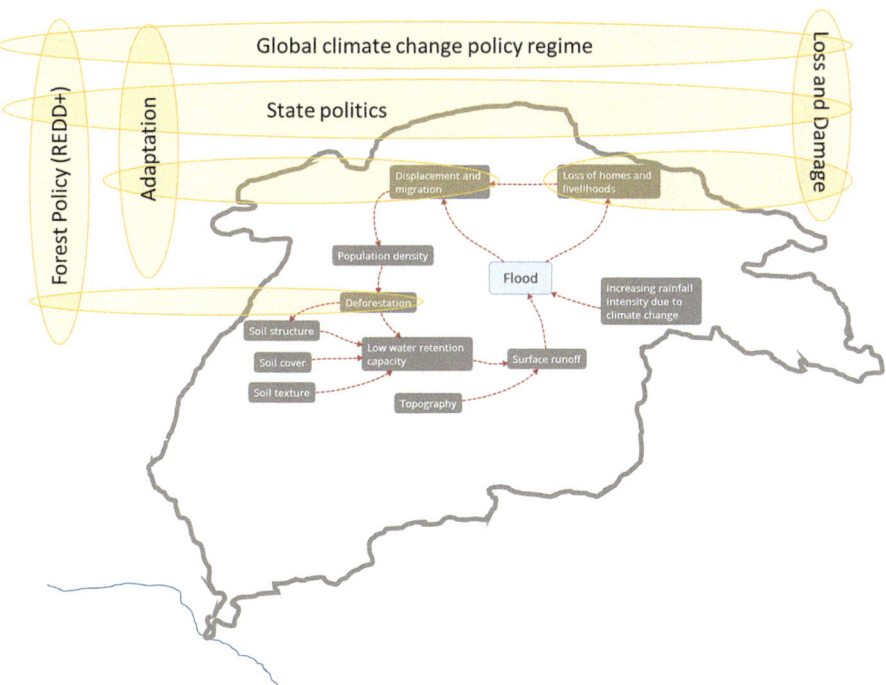

Fig. 2.1 Schematic illustration of how social fields (orange ovals) interact with natural systems (Indus river basin). (Modified from (Olsson and Jerneck 2018))

2.7   A Typology – Linking Science and Politics

Below we suggest a typology which links a scientific understanding of sustainability challenges with political and ideological beliefs. The typology is a device for reframing research problems. By shifting them between distinctly different visionary categories we make theoretical, methodological, and other features visible. This allows further scrutiny of complementarities and contradictions as well as an evaluation of the potential of these frames for tackling sustainability challenges. The typology should be seen as a source of inspiration and discussion rather than a fixed schema. Ultimately, the goal of such a typology is to increase the political awareness of scientific knowledge production as a basis for a more politically informed sustainability science which is a prerequisite for social change.

As a basis for the typology, we postulate that a spectrum of visions – one and each of which claims to promise sustainability – from continuing the ongoing modernisation to defying modernisation **exists**. Along this spectrum we define three more or less distinct views supported by their own frame of science and reality, theoretical and methodological approaches, and strategies for social change. In the typology we call them *ecological modernity, critical modernity* and *anti-modernity*. Inspired by York & Rosa's theoretical analysis of ecological modernisation theory (EMT) (York and Rosa 2003) we make a distinction between ecological modernity and critical modernity. In their analysis they distinguish between observed institutional changes and the efficacy and outcome of such changes. Proceeding from this distinction, endeavours for achieving sustainability have clearly generated a wide individual, institutional, and organisational response throughout society – at least since the time of Our Common Future. To exemplify, states participate in international negotiations resulting in national targets towards sustainability such as the Paris Agreement and the Sustainable Development Goals. Firms and corporations increasingly use sustainability claims in their communications such as Corporate Social Responsibility. Municipalities initiate and promote recycling and waste management as well as public transports. Civil society exert pressure on the private and public sectors and call for improved environmental performance while many individuals seek to adjust to these new norms. In all, these activities and processes are recognised as contributing to and providing the basis for ecological modernisation. In our view, these responses are necessary for achieving sustainability, but are they sufficient, or is there need for more?

While categories under ecological modernity focus on *promises* of social change rather than on outcomes, the categories under critical modernity are more concerned about the *outcomes*. Ecological modernisation is characterised by a set of piecemeal and incremental processes of change without any connection to a planetary whole. Critical modernisation, however, takes the global whole as its point of departure in sketching what is needed in terms of social change to achieve sustainability. Anti-modernity as a worldview is less coherent than the other two worldviews in terms of the processes of social change that may be required for sustainability. Anti-modernity has a strong focus on the, often utopian, images of sustainability (Naess and Rothenberg 1990; Taylor 2011) rather than on the processes of social change leading to these outcomes.

The boundaries between the three visions and their associated categories are fuzzy but the scheme provides a useful heuristic for understanding the rich flora of sustainability approaches and claims (Table 2.1).

Table 2.1 Three visions of sustainability

Pathway	Weak sustainability	Critical sustainability	Unclear
Theory	Ecological modernisation	Radical reform = modernisation	Anti-modernisation, degrowth/postgrowth
Problem definition	Empirically observed and approached = inductive	Empirically grounded while theoretically informed = analytic induction	Theoretically generated and approached = deductive
Analytical approach	Specific = detailed but detached	Critical reframing via several frames = varied and complex	Holistic = encompassing but vague
Main benefit	Rapid progress in solving (narrowly) identified problems	Structural change towards sustainability based on broad understanding	Visionary, activism, social movement
Main drawback	Risk of lock-in	Slow progress	Utopian
Concepts	Green state (Eckersley 2004; Taylor 2011)	Political ecology	Deep ecology (Naess and Rothenberg 1990)
Discourses	Green economy (Stern 2009)	Envisioning real utopias (Wright 2010)	De-growth (Kallis 2011)
Theory	Resilience theory (SRC 2016)	Ecological unequal exchange (Rice 2007)	Biocentric egalitarianism (Taylor 2011)
	Environmental economics	Transition theory (Geels 2011)	
		Ecological economics	
Mechanisms	Corporate social responsibility, market based schemes such as PES. Conditional cash transfers	Creative destruction and disruptive innovations, radical tax reforms.	Basic salary for all
		Unconditional cash transfers.	

2.8   The Way Forward

At this point, we challenge the notion of coupled social-ecological systems. Ontologically, we therefore separate nature, often represented by systems or models based on a system representation, and society – here represented by strategic action fields – for the purpose of creating a methodological opportunity to unite (the best available) knowledge from each in a process of integrative research.

Sustainability science has a strong focus on action-oriented research; hence, politics is essential for sustainability science. Social fields theory is a way to make the politics of sustainability visible and actionable, and by linking strategic action fields to natural systems we are able to identify the leverage points of the natural system.

To make the political dimensions visible, and to facilitate framing and reframing, we suggested a typology whereby the sustainability challenge can be placed in a spectrum of sustainability visions, from ecological modernity, through critical modernity, to anti-modernity.

References

Agrawal A (1995) Dismantling the divide between indigenous and scientific knowldege. Dev Chang 26:413–439

Barth M, Michelsen G (2013) Learning for change: an educational contribution to sustainability science. Sustain Sci 8(1):103–119

Blumer H (1986) Symbolic interactionism: perspective and method. University of California Press, Berkeley/Los Angeles

Brown K (2014) Global environmental change I: a social turn for resilience? Prog Hum Geogr 38(1):107–117. https://doi.org/10.1177/0309132513498837

Cash DW, Clark WC, Alcock F, Dickson NM, Eckley N, Guston DH, Jäger J, Mitchell RB (2003) Knowledge systems for sustainable development. PNAS 100(14):8086–8091. https://doi.org/10.1073/pnas.1231332100

Clark WC (2007) Sustainability science: a room of its own. Proc Natl Acad Sci U S A 104(6):1737–1738

Clarke S, Walsh A (2009) Scientific imperialism and the proper relations between the sciences. Int Stud Philos Sci 23(2):195–207

Dessler D (1989) What's at stake in the agent-structure debate? Int Organ 43(03):441–473

Devia GK, Ganasri BP, Dwarakish GS (2015) A review on hydrological models. Aquat Proc 4:1001–1007

Diamond J (1999) Guns, germs, and steel: the fates of human societies. W.W. Norton & Company, New York

Diamond J (2005) Collapse: how societies choose to fail or succeed. Penguin, New York

Dupre J (1991) Reflections on biology and culture. In: Sheehan JJ, Sosna M (eds) Boundaries of humanity: humans, animals, machines. University of California Press, Berkeley, pp 125–131

Dupré J (1994) Against scientific imperialism. Proc Bienn Meet Philos Sci Assoc 1994(2):374–381

Dupré J (2001) Human nature and the limits of science. Oxford University Press, Oxford

Eckersley R (2004) The Green State. Rethinking democracy and sovereignty. MIT Press, Cambridge, MA

Field CB, Barros V, Stocker TF, Qin D, Dokken DJ, Ebi KL, Mastrandrea MD, Mach KJ, Plattner GK, Allen SK (2012) Managing the risks of extreme events and disasters to advance climate change adaptation. A special report of working groups I and II of the intergovernmental panel on climate change. Cambridge University Press, Cambridge/New York

Fligstein N, McAdam D (2012) A theory of fields. Oxford University Press, New York

Gamson WA (1975) The strategy of social protest. Dorsey Press, Homewood

Geels FW (2011) The multi-level perspective on sustainability transitions: responses to seven criticisms. Environ Innov Soc Trans 1(1):24–40

Geels FW, Berkhout F, van Vuuren DP (2016) Bridging analytical approaches for low-carbon transitions. Nat Clim Chang 6:576–583

Hage J, Hollingsworth JR (2000) A strategy for the analysis of idea innovation networks and institutions. Organ Stud 21(5):971–1004

Hatt K (2013) Social attractors: a proposal to enhance "resilience thinking" about the social. Soc Nat Resour 26(1):30–43

Hay C (2002) Political analysis: a critical introduction. Palgrave Macmillan, Houndmills

Ionesco D, Mokhnacheva D, Gemenne F (2016) The atlas of environmental migration. Routledge, London/New York

Isgren E, Jerneck A, O'Byrne D (2017) Pluralism in search of sustainability: ethics, knowledge and methodology in sustainability science. Chall Sustain 5(1):2–6

Jerneck A, Olsson L (2011) Breaking out of sustainability impasses: how to apply frame analysis, reframing and transition theory to global health challenges. Environ Innov Soc Trans 1(2):255–271

Jerneck A, Olsson L, Ness B, Anderberg S, Baier M, Clark E, Hickler T, Hornborg A, Kronsell A, Lövbrand E, Persson J (2011) Structuring sustainability science. Sustain Sci 6(1):69–82

Kallis G (2011) In defence of degrowth. Ecol Econ 70(5):873–880

Kates RW (2011) What kind of a science is sustainability science? Proc Natl Acad Sci 108(49):19449–19450

Kates, R.W, W.C. Clark, R. Correll, J.M. Hall, C.C. Jaeger, I. Lowe, J.J. McCarthy, H.J. Schellnhuber, B. Bolin, N.M. Dickson, S. Faucheux, G.C. Gallopin, A. Grübler, B. Huntley, J. Jäger, N.S. Jodha, R.E. Kasperson, A. Mabogunje, P. Matson, H. Mooney, B. Moore III, T. O'Riordan, and U. Svedin. 2001. "Sustainability science." *Science* 292 (5517):641-642

Kauffman J, Arico S (2014) New directions in sustainability science: promoting integration and cooperation. Sustain Sci 9(4):413–418

Kellert SH, Longino HE, Waters CK (2006) Scientific pluralism, introduction. Minnesota studies in the philosophy of science, vol XIX. University of Minnesota Press, Minneapolis

Kumar P (2010) The economics of ecosystems and biodiversity: ecological and economic foundations. UNEP and Earthscan, London/Washington, DC

Lang DJ, Wiek A, Bergmann M, Stauffacher M, Martens P, Moll P, Swilling M, Thomas CJ (2012) Transdisciplinary research in sustainability science: practice, principles, and challenges. Sustain Sci 7(1):25–43

Levitt P, Schiller NG (2004) Conceptualizing simultaneity: a transnational social field perspective on society. Int Migr Rev 38(3):1002–1039

Mahoney J (2000) Path dependence in historical sociology. Theory Soc 29:507–548

Mahoney J, Thelen K (2010) A theory of gradual institutional change. In: Mahoney J, Thelen K (eds) Explaining institutional change: ambiguity, agency, and power. Cambridge University Press, New York, pp 1–37

Mäki U (2013) Scientific imperialism: difficulties in definition, identification, and assessment. Int Stud Philos Sci 27(3):325–339

Martin JL (2003) What is field theory? 1. Am J Sociol 109(1):1–49

Meadowcroft J (2011) Engaging with the politics of sustainability transitions. Environ Innov Soc Trans 1(1):70–75

Midgley M (1984) Reductivism, fatalism and sociobiology. J Appl Philos 1(1):107–114

Miller TR, Wiek A, Sarewitz D, Robinson J, Olsson L, Kriebel D, Loorbach D (2013) The future of sustainability science: a solutions-oriented research agenda. Sustain Sci 8(3):1–8. https://doi.org/10.1007/s11625-013-0224-6

Mitchell SD (2009) Unsimple truths: science, complexity, and policy. University of Chicago Press, Chicago

Naess A, Rothenberg D (1990) Ecology, community, and lifestyle: outline of an ecosophy. Cambridge University Press, Cambridge

Newton AC (2016) Biodiversity risks of adopting resilience as a policy goal. Conserv Lett 9(5):369–376

Norgaard RB (1989) The case for methodological pluralism. Ecol Econ 1(1):37–57

Olsson L, Jerneck A (2018) Social fields and natural systems: integrating knowledge about society and nature. Ecol Soc 23(3):1–18. https://doi.org/10.5751/ES-10333-230326

Olsson L, Opondo M, Tschakert P, Agrawal A, Eriksen SH, Ma S, Perch LN, Zakieldeen SA (2014) Livelihoods and poverty. In: Field CB, Barros VR, Dokken DJ, Mach KJ, Mastrandrea MD, Bilir TE, Chatterjee M, Ebi KL, Estrada YO, Genova RC, Girma B, Kissel ES, Levy AN, MacCracken S, Mastrandrea PR, White LL (eds) Climate change 2014: impacts, adaptation, and vulnerability. Part A: global and sectoral aspects. Contribution of working group II to the fifth assessment report of the IPCC. Cambridge University Press, Cambridge/New York, pp 793–832

Olsson L, Jerneck A, Thoren H, Persson J, O'Byrne D (2015) Why resilience is unappealing to social science: theoretical and empirical investigations of the scientific use of resilience. Sci Adv 1(4):e1400217

Parsons M, Nalau J, Fischer K (2017) Alternative perspectives on sustainability: indigenous knowledge and methodologies. Chall Sustain 5(1):7–14

Persson J, Hornborg A, Olsson L, Thoren H (2018a) Toward an alternative dialogue between the social and natural sciences. Ecol Soc 23(3)

Persson J, Johansson E, Olsson L (2018b) Harnessing local knowledge for scientific knowledge production: challenges and pitfalls within evidence-based sustainability studies. Ecol Soc 23(3)

Rice J (2007) Ecological unequal exchange: international trade and uneven utilization of environmental space in the world system. Soc Forces 85(3):1369–1392

Ritzer G (2011) Sociological theory. McGraw-Hill, New York

Sachs JD, Warner AM (1995) Natural resource abundance and economic growth. National Bureau of Economic Research, Cambridge, MA

Sala S, Farioli F, Zamagni A (2013) Progress in sustainability science: lessons learnt from current methodologies for sustainability assessment: part 1. Int J Life Cycle Assess 18(9):1653–1672

Salas-Zapata WA, Rios-Osorio LA, Trouchon-Osorio AL (2013) Typology of scientific reflections needed for sustainability science development. Sustain Sci 8(4):607–612

Scott WR (1995) Institutions and organizations. Sage, Thousand Oaks

Shahadu H (2016) Towards an umbrella science of sustainability. Sustain Sci 11(5):1–12

Siedlok F, Hibbert P (2014) The organization of interdisciplinary research: modes, drivers and barriers. Int J Manag Rev 16(2):194–210

Sluyter A (2003) Neo-environmental determinism, intellectual damage control, and nature/society science. Antipode 35(4):813–817

Spangenberg JH (2011) Sustainability science: a review, an analysis and some empirical lessons. Environ Conserv 38(3):275–287

SRC (2016) What is resilence? An introduction to a popular concept. Stockholm Resilience Centre. Accessed 3 Dec. http://www.stockholmresilience.org/research/research-news/2015-02-19-what-is-resilience.html

Stern N (2009) A blueprint for a safer planet: how to manage climate change and create of a new era of progress and prosperity. Bodley Head, London

Taylor PW (2011) Respect for nature: a theory of environmental ethics. Princeton University Press

Thoren H (2015) The hammer and the nail. PhD, Lund University, Department of Philosophy van Kerkhoff L (2014) Developing integrative research for sustainability science through a complexity principles-based approach. Sustain Sci 9(2):143–155

Walker BH, Gunderson LH, Kinzig AP, Folke C, Carpenter SR, Schultz L (2006) A handful of heuristics and some propositions for understanding resilience in social-ecological systems. Ecol Soc 11(1):13

Wellstead A, Howlett M, Rayner J (2016) Structural-functionalism redux: adaptation to climate change and the challenge of a science-driven policy agenda. Crit Pol Stud 11:1–20

Wiek A, Ness B, Schweizer-Ries P, Brand FS, Farioli F (2012) From complex systems analysis to transformational change: a comparative appraisal of sustainability science projects. Sustain Sci 7(1):5–24

Wright EO (2010) Envisioning real utopias, vol 98. Verso, London

York R, Rosa EA (2003) Key challenges to ecological modernization theory institutional efficacy, case study evidence, units of analysis, and the pace of eco-efficiency. Organ Environ 16(3):273–288

Chapter 3
Approaches for Framing Sustainability Challenges: Experiences from Swedish Sustainability Science Education

Barry Ness

Abstract Sustainability challenges are defined by their complex and multifaceted interactions between nature and society and contention as to how and where to direct problem-solving efforts. This chapter presents four different approaches that exist for framing sustainability challenge areas that are introduced and worked with by students in LUMES International Master Programme in Environmental Studies and Sustainability Science at Lund University in Sweden. The approaches include the (1) Driver-Pressure-State-Impact-Response (DPSIR) framework, (2) causal loop diagrams (CLDs), (3) multi-scale and level perspective, including transition theory and management, and the SES framework. Each approach is described and critically assessed, especially from the perspective of student mastery. The outcome of the chapter is a more comprehensive understanding of which approaches are useful for different sustainability problem constellations and a deeper comprehension of how the framing tools can be taught in sustainability science education.

Keywords Framing approaches · Sustainability education · DPSIR · CLDs · Transition theory · SES framework · Sweden

3.1 Introduction

The field of sustainability science (Kates et al. 2001; Ness 2013; Jerneck et al. 2011) has experienced rapid development since the turn of the twenty-first century. The advances have extended down multiple trajectories within the realms of research and education. One area where ambitions have been strongest is with efforts to more closely link scholars to knowledge creation and problem-solving processes outside of academia (Wiek et al. 2012; Spandenberg 2011). Many of the recent developments—with aspirations to guide societal change along more sustainable trajectories—have been carried out through a diverse set of transdisciplinary and transformative methods with diverse actors through unique processes as *transition*

B. Ness (✉)
Centre for Sustainability Studies (LUCSUS), Lund University, Lund, Sweden
e-mail: barry.ness@lucsus.lu.se

© The Author(s) 2020   35
T. Mino, S. Kudo (eds.), *Framing in Sustainability Science*,
Science for Sustainable Societies, https://doi.org/10.1007/978-981-13-9061-6_3

experiments or *(urban) living labs* (Nevens et al. 2013; Evans and Karvonen 2014; Baccarne et al. 2016; Buhr et al. 2016). However, before efforts to address targeted sustainability challenges can take place, it is common for a robust and preferably unified understanding—or *framing*—of problem areas to occur (Ness et al. 2010). How different actors carry this out can vary greatly. A number of conceptualization approaches have been developed, or adopted from other disciplines and fields, for the purpose of better comprehending coupled socio-ecological systems. They have been developed around the perspective of shared boundary concepts (e.g. resilience, vulnerability, ecosystem services), common objects (e.g. maps), a common theoretical perspective, or defined (sustainable) development priorities (Cash et al. 2003; Clark et al. 2016).

3.1.1 Education for Sustainability

For framings to be salient and robust, proficiencies to derive common problem conceptualizations must be developed amongst scholars, facilitators, and other actors. One important forum for fostering these skills is in university sustainability education programs. Although skills training in this area traditionally has been beyond the scope of most educational programs, where the focus is usually on *descriptive/analytical* modes of performing research, a number of sustainability programs have recently been established—or redeveloped—under the umbrella of transformative education (Schneidewind et al. 2016). The curricula in these programs respond to priorities that participants are *not only* able to analyze sustainability problems and suggest solutions; the education also empowers them to become agents of (sustainable) change, to predict and prepare for new challenges, and to create new opportunities to infuse sustainability into societal processes at different scales and levels. Focus and student proficiency development of these areas has been devised with an explicit focus on multiple and often competing comprehensions in sustainability problem areas as well as where solutions can be directed and experimented with. To operationalize these, a number of the programs have been augmented to include student development of *key competencies* for future researchers and sustainability practitioners (Wiek et al. 2012; Wiek and Kay 2015; Burns 2015). One prominent set of competencies developed by Wiek et al. (2011) include *systems thinking, strategic, anticipatory, normative* and *interpersonal* abilities. Focusing on these five areas creates opportunities for students to gain proficiencies and expertise in areas such as future visioning and scenarios, systems analysis, ethics, risk, and group facilitation to name just a few.

One useful educational forum to foster the competency development—especially concerning framing—is group work. Group work allows many students to gain an understanding that they almost certainly would not have developed individually, fostering reflexivity amongst participants where broader worldviews are exchanged, reflected on, challenged, and compromised on within the hopeful *safespace* of trust and understanding amongst participants. Furthermore, the activities allow for a *division of labor* amongst students as an approach to managing complexity and the multifaceted nature of sustainability challenges.

3.1.2 Aims

Many approaches can be used in a participatory manner to frame sustainability challenges and help expose potential solution options for the challenges. Some approaches have been developed specifically for certain challenges; others are broad approaches that are useful for encapsulating the dynamics of a variety of systems or questions. Despite their existence and analyses of them, insufficient understanding remains as to which approaches are useful for which framing and problem assessment purposes. Furthermore, inadequate attention has been paid to how to best nurture student competencies in using the different approaches, especially training in settings where actors differ in societal facets.

This chapter presents and critically assesses four approaches for framing and structuring sustainability challenges. The assessment is conducted from how each is used by students in the sustainability science course of the Lund University International Master's Program in Environmental Studies and Sustainability Science (LUMES) in Sweden. It includes a set of approaches that can be applied to diverse sustainability challenges and is based on broader concepts of causality or scale. The approaches presented are the DPSIR framework, causal-loop perspective in transition theory, and the socio-ecological system framework because they are robust and commonly found in the sustainability literature. This review provides reflections and insights from both the perspective of student learning of the approaches and perceptions on how the approaches can be taught to foster student skills development, particularly in a limited time frame. Each approach is described and critically assessed, especially from the context of student learning activities. The outcome of the chapter is a more comprehensive understanding of the four approaches, including their respective strengths and weaknesses. In addition, there are insights on how student proficiency in using the approaches can be fostered. The main empirical material used in the study is course evaluations from course participants over the past 7 years (2011–2018), notes from face-to-face group follow-up course evaluation sessions, and where available, instructor reflection notes on individual student learning activities.

The chapter is structured as follows. First, the LUMES graduate program and more specifically the sustainability science course is presented. Next, a differentiation between the diverse terminologies used when describing the approaches is completed, followed by a presentation of each of the framing approaches. Subsequently, the possibilities, limitations, along with insights from student learning perspectives are carried out. The chapter concludes with a discussion covering what has been done in recent years to improve the student learning processes and general reflections on student key competency development.

3.2   The LUMES Program

3.2.1   Program Structure

The LUMES Program (Lund University International Master Program in Environmental studies and Sustainability Science) is a 2-year graduate program with approximately 40–50 students annually. Participants are from diverse academic backgrounds and nationalities. The program was launched in 1997 and has undergone two major curriculum redevelopment processes. It is a cohesive program where students take all first-year courses together as a single group. The program consists of three 10-credit core courses during the first term: earth system science, social theory, and sustainability science. During the second term, students take a number of broader thematic courses including governance for sustainability, urban and rural systems, economy and sustainability. In addition, there is one extended course, knowledge to action, which spans part of the first term and the entire second term. This course has strong ties to the sustainability science course. During the third term, students must successfully complete four of a variety of targeted courses offered during the term: energy, water, global health, gender, and social movements amongst numerous others. Students complete the program with the successful submission, presentation, and defense of a Master's thesis on a sustainability-related topic that they design individually and conduct research.

3.2.2   Sustainability Science Course

The LUMES Sustainability Science course is one of the three main courses of the first term of the program. It acts as a bridge to link the initial two courses, which greatly differ from each other. The course runs from late-November until late-January with the holiday break of around 2 weeks in the middle. The course has strong topical connections and schedule overlap with the knowledge to action course. Learning outcomes for the sustainability science course—in differing manners and degrees—center on the key competency areas with concentrated student knowledge and skills development efforts on the history and evolution of sustainability science, the main concepts in the field, (e.g., systems thinking, complexity, socio-ecological systems, inter- and transdisciplinarity, resilience, political ecology, transitions), interpersonal skills through multiple presentations, and group work activities. There also is training in anticipatory competencies via a short learning segment on scenarios and envisioning. In addition, and covering multiple competency areas, there is an emphasis on student comprehension of the different framing approaches with a strong focus on the applicability, strengths, and weaknesses, of each of them. The course structure is varied with learning activities on the development of sustainability science, broader systems thinking/tools for measuring sustainability, and a block on inter- and transdisciplinary sustainability research. Students are evaluated both individually and by groups. Individual assessment is carried out via targeted reflection assignments (e.g., literature reflection, systems thinking reflection); group assessment takes place through a collection of presentations, group reflection papers, and a final project report and presentation.

Depending on the approach, there are roughly 2 days devoted to each. Each approach block is supplemented with student reading of two to five scholarly articles, which students are instructed to read in advance. For each block there is a 1- to 2-h lecture by the course instructor explaining, for example, its developmental history, application, and examples of how and where the approach has been applied. Augmenting the lecture and readings, there also can be a presentation from an "expert" from outside of LUCSUS (Lund University Centre for Sustainability Studies) with greater research and/or practical experience with the specific approach.

For students to develop a greater understanding and increased competency levels with the approaches, learning activities for each are performed in smaller, randomly generated groups of five-six students. In these groups, students are paired to an ongoing—or desired—research topic carried out by an early-career researcher and project mentor based at LUCSUS. The researcher is responsible for ensuring that students receive an overview of the general topic, targeted topic advice, and/or basic readings on the theme. Final student group topics have varied greatly, focusing on, for example, coastal management in Florida, food security and production systems in Uganda, mangrove destruction from biofuel feedstock production in Indonesia, land grabbing in Tanzania, and bush meat production and trade in Ecuador, to name a few. Each group then concentrates and develops their respective topic as each new framework is presented to the entire class. However, one exception is the social ecological system framework where experience has demonstrated that performing a sufficient assessment using the approach for each topic would take far too long in the limited time available during the course. Instead, students work on one common case where each group concentrates on a particular subsystem (e.g., governance system, resource units).

Important to the student comprehension of each approach and the broader project is the respective group's formulation of an appropriate focus/question and definition of "boundaries" (i.e., what parameters are included, what is left out). Collectively—and often in an iterative fashion—the group then devises a conceptualization (model) by using the approach to address the specific question posed. The groups then work through several iterations of a framing while receiving constructive feedback from one another and the respective topic mentor. For each approach course block, the groups also present their respective conceptualizations to classmates and the instructor in a small seminar session. This provides opportunities for students to learn through what others have done, and to gain additional insights on their own work through fellow student and instructor critique and feedback.

Additionally, there is a final project deliverable where each group combines a number of the approach conceptualizations (e.g., CLD and transition theory multilevel perspective, SES and DPSIR) in a hopefully coherent "package" based on a specific topic aim or question that they unify around. Through the lens, they then reflect more deeply on each approach employed (e.g., strengths, limitations), and on the broader context of framing complex sustainability challenges. Three examples of 2016–2017 projects included palm oil production transition using the multilevel perspective and a DPSIR scheme, barriers to the change to an organic viticulture system in California using CLDs and the SES framework, and small scale hydropower development in Nepal, also using the multilevel perspective and DPSIR. Student group work is presented in a final seminar at the end of the course where again the project is scrutinized by classmates, mentors, and the instructor. Furthermore, students deliver a final written group project summary of roughly eight pages text that is evaluated by the course instructor. The written summary helps students to further develop writing proficiencies, especially in the area of concise writing.

3.3   The Approaches

3.3.1   Terminology

The terminology used to describe each framing approach can differ. Those presented here go by a number of names including *tools*, *frameworks*, *schemes*, and *techniques* all with modest epistemological and definitional differences. In this chapter, *approach* is used as an encompassing term. Where appropriate, I also use the name that each is most often referred to in academic literature. *Schemes* are systematic or organized configurations of correlated things; whereas *tools* are purposive, used as a means of accomplishing some sort of assessment task. Nobel Laureate Elinor Ostrom (Ostrom 2011) provides some differentiation between the different descriptive terms used in a hierarchical manner. She describes a *framework* as a meta-language, or *metatheoretical map* (Ostrom and Cox 2010), denoting a generalized form of theoretical analysis. *Theories* (e.g. transition theory, rational choice), on the other hand, are the working assumptions and hypothesized specifications of the framework variables deemed sufficient to provide adequate explanations or diagnoses of social and/or ecological conditions. Related to the above, *models* use more targeted assumptions about variables, predictions about the results of combining these variables using a particular theory.

3.3.2   DPSIR

DPSIR is an analysis scheme for describing cause-effect relationships in connection with environmental and natural resource management challenges (Bowen and Riley 2003; EEA 1999; Giupponi 2007). DPSIR stands for *Driving forces-Pressure-State-Impact-Response*; the scheme has been associated significantly with the European Environment Agency in Copenhagen, Denmark. The intention and the strength of the DPSIR scheme are its ability for practitioners to gain an overview of targeted (environmental) policy issues, and to estimate the appropriateness and efficiency of different governance responses (EEA 1999). It also permits the integration of socio-economic and ecological system information into one framework (Bidone and Lacerda 2004). The scheme helps to structure information into the five distinct areas, making it possible to identify and structure the important causal relationships. DPSIR conceptualizations can be simplistic or sophisticated dependent on the focus and/or the question(s) they address. The scheme has been used extensively for challenges to water and coastal regions (Gari et al. 2015). Figure 3.1 represents a simple depiction of the DPSIR framework for Baltic Sea eutrophication from Swedish agriculture.

The DPSIR approach has evolved from a long line of more simplistic frameworks for environmental issues such as Statistics Canada's Stress-Response (S-R) framework from the late 1970s (Gari et al. 2015), the Pressure-State-Response (P-S-R) scheme launched by the Organization for Economic Cooperation and Development in the 1980s, and the United Nations Commissions on Sustainable Development's Drivers-Pressure-Response (D-P-R) framework (OECD 1994).

The DPSIR approach has received considerable critique as well; it has often been directed at the mechanistic nature and oversimplification of the scheme, its linearity, and the difficulty in handling parameters that may be a part of multiple DPSIR phases (e.g., driver and state conditions) (Klijn 2014). An additional challenge is with its ability in incorporating the multi-dimensional and multi-scalar causal relationships of problems where many sustainability issues are characterized by complex dynamics in time and space are worsened by multiple and interacting anthropogenic and natural driving forces (Kates et al. 2001). These issues include,

Fig. 3.1 Simple DPSIR for Baltic Sea eutrophication from Swedish agricultural production (Ness et al. 2010)

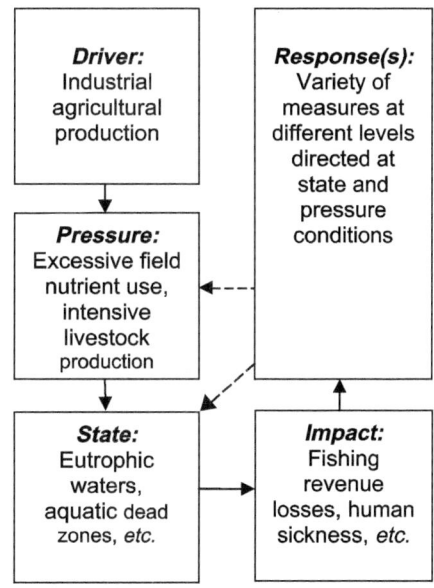

for example, global climate change, poverty, eutrophication, and biotic diversity. Finally, the DPSIR framework has historically been developed and used for presenting environmental impacts caused by socio-economic driving forces. Analyses of socio-economic system state conditions and impacts (e.g. HIV/AIDS, malaria, and poverty) have seldom been included in such analyses—thusly not reflecting the broad variety of sustainability challenges (Ness et al. 2010). To address many of the deficiencies along with making the scheme more useful for targeted areas, DPSIR has continued to be developed and augmented by scholars and practitioners to include, amongst numerous others, the 'EBM-DPSER' concentrating on ecosystem services (Kelble et al. 2013), the 'DPSWR' on human welfare (O'Higgins et al. 2014), the 'eDPSEEA' for Health (Reis et al. 2015), and the multi-level DPSIR (Ness et al. 2010).

3.3.3   Causal Loop Diagrams

A causal loop diagram (CLD) is a general approach to the qualitative analysis of systems; CLDs incorporate both human and social parameters into a single, sometimes sophisticated, conceptualization. They are often used as a part of a broader participatory systems analysis approach, including problem and system boundary definition, qualitative conceptualization creation, and quantitative system dynamics modeling. A strength of CLDs is that they are a flexible framework where creators identify and describe, in increasing levels of complexity, the cause-effect relationships of different sub-components of a larger system. Arrows are used to link cause-effect relationships, connecting the two components.

The diagrams use different symbols to denote different relationships. A positive plus [+] symbol between two variables indicates a parallel behavior of the two, meaning an increase in the causative variable also causes the effect variable to increase; furthermore, a decrease in the causative variable denotes a decrease in the affected variable. Conversely, a negative minus [−] symbol indicates an inverse relationship between the two variables, meaning as the causative variable increases, the affected variable decreases, or *vice-versa*. Numerous sub-components of a system can form loops, feeding back on one another, either directly or indirectly. A loop that has a reinforcing behavior is often denoted in the diagram with 'R'; this signifies exponential growth of that subsystem. Loops denoted with 'B' indicate a balancing behavior of the subsystem. Temporal aspects in the form of time lags can also be identified in the CLD using two parallel lines through the center of the arrow linking the variables. An example of a simplistic CLD for bush encroachment in southern Africa is shown in Fig. 3.2 (SAPECS 2016). The arrangement shows the causal relationships of two drivers of global climate change and human population growth in the region and their ultimate impacts on such factors as woody plant growth, land area and water availability.

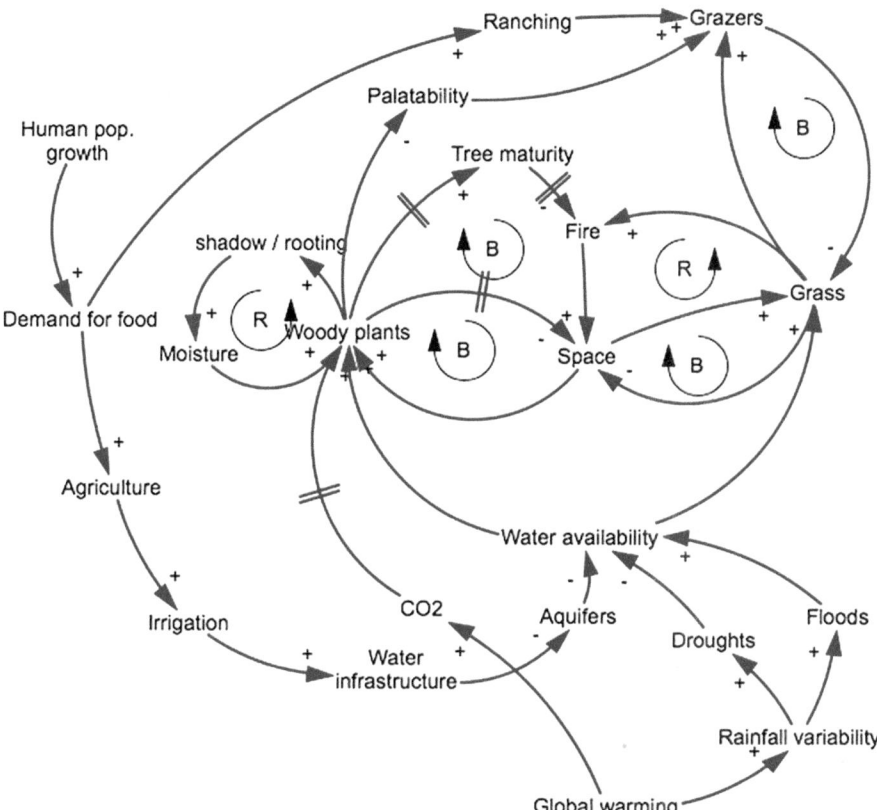

Fig. 3.2 Example of a simple causal loop diagram for bush encroachment in southern Africa. The conceptualization shows the main drivers of the encroachment and their causal impacts on other parameters. (Source: Southern African Program on Ecosystem Change and Society, n.d.)

CLDs are a useful approach for grasping the casual interactions of defined systems and like the DPSIR scheme, allow the practitioner to experiment with solutions to the particular challenge area. However, CLDs possess a number of shortcomings that can influence their usefulness in framing sustainability challenges. First, the labeling of the different sub-components can appear problematic. The parameters must always be labeled as more or less of something (e.g., human population, greenhouse gas releases, biodiversity loss). This can lead to difficulties in understanding the respective sub-components of a system. In addition, critique has been lodged against a CLD's *spaghetti-like* appearance, and related inability in understanding sophisticated conceptualizations of a problem area. Related, the aim of a CLD is to create causal relationships in as few steps as possible. Gross oversimplifications in processes also can often cause difficulties in interpreting a CLD therefore creating opportunities for creating false conclusions to be drawn about the system in question.

3.3.4   Multi-scale & -level Perspective (Including Transitions)

Another approach for understanding and structuring sustainability challenges is through the multi-scale and -level perspective. This form of assessment has been promoted and used for decades, and has been used for a variety of socio-ecological systems including sustainable tourism (Crnogaj et al. 2014), wastewater treatment systems (Molinos-Senante et al. 2014), water resources management (Daniell et al. 2014), climate change (Bulkeley and Betsill 2013), and renewable energy transformations (Di Lucia and Ericsson 2014), to name a few. *Scale* refers to the analytical dimensions for measuring and studying objects and processes. Examples of different scales can be spatial, administrative, jurisdictional, managerial, or temporal. *Levels* refer to locations along those scales (Gibson et al. 2000). Related to these is hierarchy. A *hierarchy* is a conceptually linked system for grouping phenomena along a particular scale.

The strength of the approach is not based on causal relationships between phenomena as with the initial two approaches; instead, applying the perspective creates the ability to match usually distinct bio-geo-physical systems scales with social system scales such as management systems (Cash and Moser 2000) where the practitioner gains a robust understanding of a problem constellation. Like the first two approaches described, conceptualizations can be simple or sophisticated depending on the phenomena assessed. Additionally, an important intention with this approach is to detect where *disconnects* or *mismatches* can lie between different scales or levels (Cash and Moser 2000).

Scales can be predominantly inclusive or exclusive (Gibson et al. 2000). An inclusive (or nested) hierarchy is a group of objects or processes that is contained in subdivisions of groups of higher systems such as the modern taxonomic classification. An exclusive hierarchy is where groups of objects (or processes) in a lower ranked hierarchy are not included or as subdivisions of higher ranked groups such as the military ranking system (Gibson et al. 2000).

3.3.5   Multi-level Perspective in Transition Theory

A particular type of approach for understanding processes of sustainable change, often over time, is the multi-level perspective (MLP) in transition theory and management. Broadly, transitions are deliberate processes of societal change in culture, practices and structure (e.g., agroecology in Uganda, renewable energy development in Sweden) (Nevens et al. 2013; Geels 2011). This mid-level theory is an extension of socio-technical systems rooted in sociology, institutional theory, and innovation studies (Geels 2004). Studies in this research field examine complex adaptive systems from the perspectives of long-term processes and non-linearity (Avelino and Rotmans 2009). The objects of focus of transitions are not abrupt, fast societal (sustainable) change; rather, a transition is an incremental and constant process of change where the fundamental character of society—or a sub-system of society—transforms (Rotmans et al. 2001). The field has extended to sustainability over the past decade-plus with a number of "experiments," especially in urban areas throughout Europe. A conceptualization of the three levels with more specific divisions of different socio-technical regimes and how a niche can emerge over time is displayed in Fig. 3.3.

The MLP in transitions consists of three unique levels to encapsulate the social dynamics: *landscape, regime,* and *niche.* Landscape development (macro-level) refers to the broad societal material and immaterial elements. These landscape are the important elements that "surround" the particular system of study (Avelino and Rotmans 2009). Examples can include public infrastructure or concepts that dominate societal discourses (e.g. sustainable development, resilience, free-market economy). Regimes are patchworks of institutions and actors that support the societal *status quo* (Avelino and Rotmans 2009); they represent the rules that set the boundaries private action and public policies (Rotmans et al. 2001; Hägerstrand 2001) Finally, niches are small areas of experimentation, innovation, and learning that challenge the stability of socio-technical regimes. They are often protected spaces to deviate from the regime, and, if successful, eventually become a regime themselves (Geels 2004).

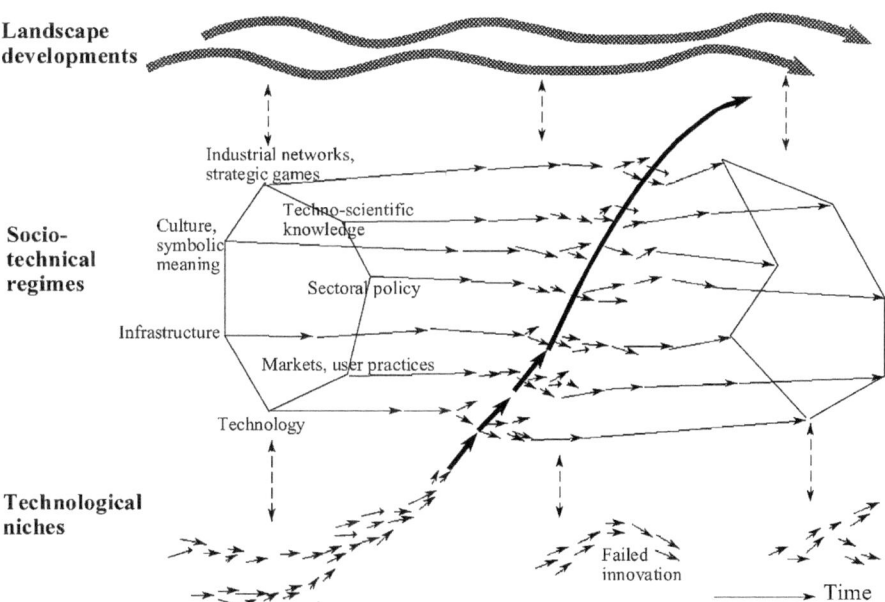

Fig. 3.3 Example of the main levels and parameters in a multi-level perspective in transition theory. The conceptualization shows the interplay play between socio-technical regimes, consisting of a number of societal institutions, and the landscape and niche levels. (Source: Geels 2004)

3.4 Socio-ecological System Framework

Another multi-level perspective approach for framing complex problems is the socio-ecological system (SES) framework. The approach has strong connections to the institutional analysis framework (IAD) (Ostrom 1990) work by Elinor Ostrom and colleagues to combine and to better understand the interactions and subsequent outcomes of complex social phenomena and ecological systems.

This classificatory framework is useful for how actors self-organize around the use of common pool resources, and around the identification of strategies to safeguard the resource in question (Ostrom 2007, 2009). A conceptualization, often constructed in a participatory manner, contributes to identifying common and relevant variables for a specific resource system. The strength of the framework is its capability for users to connect a number of multilevel nested systems. The core subsystems are the resource system, resource units, resource users, and the resource's governance system; each of these is influenced by related social, economic, and political settings as well as related ecosystems (Fig. 3.4). Each core subsystem consists of a number of examples of second-level variables that can be categorized; relevancy of each variable depends on the system in focus. Examples of the variables include size of resource system, economic value of the resources units, property-rights systems in place, and the history of use of the resource to name a few (Ostrom 2009). The intended outcome from using the framework is to devise a common set of hopefully relevant variables and sub-variables for further analysis

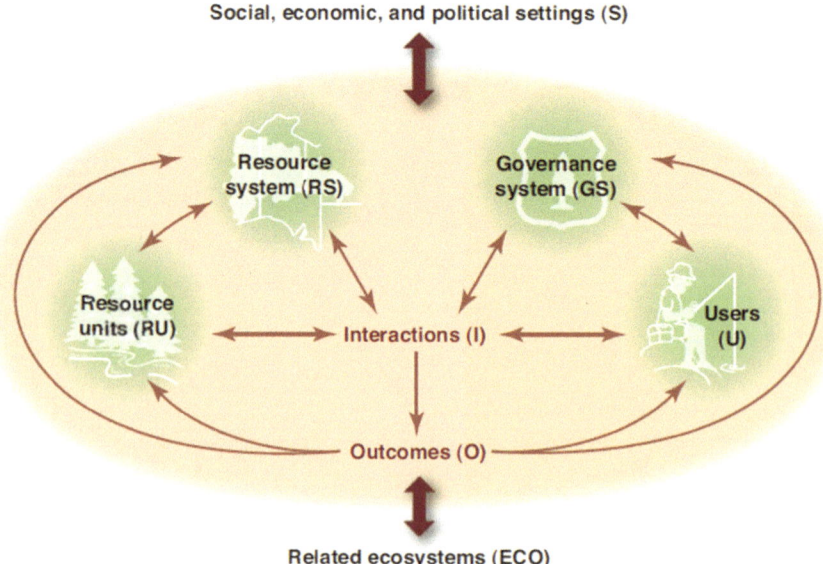

Fig. 3.4 Main subsystem interactions in the SES framework. Each subsystem also consists of a set of more targeted indicators that can be used to a more nuanced analysis (Ostrom 2009)

(e.g., data collection design, fieldwork, and analysis) for the common-pool resource (Ostrom 2009).

The SES framework, and earlier renditions of it have been applies to a variety of cases, both common-pool and non-common-pool resources. The cases include lobster fisheries in southern California (Partelow and Boda 2015), wetlands in the northern Sierra Nevada foothill oak woodlands (Hruska et al. 2014), Cambodian cattle-owning smallholders (Marshall 2015), and small-scale fisheries in Baja California Sur, Mexico (Leslie et al. 2015).

The intention of the SES framework is to undergo continuous development based on different shortcomings and case examples (Ostrom 2009). Framework development has been carried out on a number of areas where, for example, more relevant variables have been added for the case of Pacific lobster fisheries (Partelow and Boda 2015), a change in the attributes of governance systems, and ways to make the framework applicable to policy settings beyond natural resources (McGinnis and Ostrom 2014).

3.5 Discussion

3.5.1 Approach Learning Challenges

Responses from course evaluations and from classroom course evaluation sessions following the course have showed a general and diverse mix of student satisfaction around learning the approaches, as well as learning activities that need improvement. There was general displeasure, especially a number of years back, in two related areas. The first was with problems achieving a sufficient level of understanding with each approach introduced. Many of the comments from students concentrated on the lack of time and opportunities during the course to optimally learn the fundamentals of each approach. The second significant area of dissatisfaction was the existence of a common *thread* running through the entire course and the difficulty of students to see each approach in a broader perspective of sustainability science, frameworks, and tools.

More recently, and with several changes to the course learning activities, student evaluation comments have become more concentrated on single approaches with learning activity suggestions based on the individual styles of learning preferred by individual students. Course evaluation comments do not identify any single approach, lecture, or group work activity as problematic. Instead, there is a diverse mix of both positive and critical comments in all areas. This is interpreted as positive given the large size of the course and the diversity of cultural and academic backgrounds. As examples, a number of participants have been critical of the high degree of group work activities—and the activities that are graded in groups—throughout the course. Others, however, expressed the ongoing participatory knowledge creation processes as highly positive and the forums were where the most skills and competencies were fostered. Additional critical comments also were often centered on unclear instructions for the exercise, or the applicability, or difficulty, of applying a particular approach to the individual case that was assigned to the group where —depending on the questions posed by students—certain approaches just have a more natural fit with particular topics. Finally, another ongoing challenge voiced by students has concentrated on the reading materials used as a backdrop to each approach block, especially comments of the articles containing an insufficient amount of case examples.

3.5.2 Changes to Enhance Approach Understanding

A common challenge—which is not unique to the pedagogical challenges here—is fostering student depth and mastery of each approach in a (very) limited amount of time. The challenge is augmented when the student comprehension and mastery of the approaches is for 35–45 students with diverse cultural and academic backgrounds. The difficulty is also compounded by conscious efforts to create tangible connections to the earth system science and social theory courses.

Based on the feedback from students, an important characteristic of Swedish education, several changes to the course have been made to foster increased student comprehension of the framing approaches, and to create a more coherent structure throughout the entire course. With the changes, or small *tweaks*, the learning activities for the individual framing approaches have become progressively better, especially in recent years. A few of the main changes are presented here.

3.5.3 Single Case

A common critique in the past was the disparate nature of the course, especially related to the approaches. To add coherency, single group topics (with mentors) were introduced. Although originally intended to better link student learning activities to actual research taking place at LUCSUS, an ongoing recommendation of students in past years, the introduction of the topics has helped to nurture a connection to the respective mentors—albeit for only a brief period. Furthermore, they have provided an effective medium to test each framing approach in real-world sustainability problem research contexts. It has also been an important approach to create opportunities for students to see the possibilities and related pitfalls for each approach introduced. Related, the single case has also fostered increased depth in understanding with each of the approaches (or combinations of them) through an implicit object of focus of the particular case (e.g., understanding processes of change over time, governance dynamics of a system, complex causal interactions). Because of these reasons, the students have been positive about the concentration of the individual topic throughout most of the approach learning activities.

Despite the added value of the single cases, they have not been without challenges. With the introduction of staff research topics, there has, at times, been the excessive group concentration on the themes themselves (e.g., targeted problems, geographic region, potential solutions) and an insufficient focus and greater reflection on the approaches themselves. Groups have placed excessive amounts of time on the details of their respective cases, and significantly less time on learning the suitability, strengths, limitations, and weaknesses of the individual approach, or understanding each in a broader context. One additional measure to keep the focus adequately on the framing approaches has been to inform the respective mentors of the objectives of the learning activities.

3.5.4 Learning Activity Streamlining

One simple way of reducing the intensity of the course is reducing the amount of content introduced, especially the sheer number of framing approaches. Although there has been a limited amount of content streamlining over recent years largely because of ulterior reasons (e.g. resilience, removal of systems dynamics). However, when surveyed, students have been strongly opposed to the further removal of framing approaches covered during the course. Instead, efforts have concentrated on fostering greater efficiency within individual learning activities, advancing both comprehension depth and learning activity diversity for each approach. Measures taken include schedule changes to enable students to have sufficient time to read the literature in advance of the respective learning activities; the addition of extended, single-day learning sessions including both lectures and group work, varied learning activities (e.g. World Cafés, role plays), and clearer communication of expected objectives and outcomes to students. Course evaluations have shown that the changes have greatly improved satisfaction levels amongst sustainability science course participants.

3.5.5 Reflection Sessions

In addition to the two areas described above, and to create more approach coherency and generate deeper reflections for each framing approach, there has also been additional scheduled reflection sessions added to the end of each approach-learning block. The class discussions and reflections last for 15- to 45-min depending on the approach. The sessions are a forum for students to reflect deeper on the approaches and pose questions to the instructor, and one another. The sessions also serve as an opportunity to introduce the next approach to be covered and introduce the readings for the subsequent discussion.

3.5.6   Approach Readings

One challenge in teaching the individual approaches has been the assigned reading for each. More comprehensive textbooks in sustainability science have only recently started to appear, including this one. However, because of the unique collection of approaches taught in the course, no single textbook is adequate. This warrants the use of scholarly articles for each approach. A challenge has been to find readings that provide an adequate overview for each approach, are not repetitive, and hopefully also provide a case example of how the approach can be applied. This challenge has resulted in the continual updating of preparatory reading materials for each framing approach, often informed by student reflections of each reading. The adequacy of the readings, however, is expected to increase in the future as additional articles—especially case examples (e.g. multi-level perspective, SES framework)—become available in the academic literature.

3.5.7   Final Reflections

Teaching and working with LUMES students over the past years generating competencies with framing complex sustainability challenges has been challenging. Simultaneously, it has also been one of the most fulfilling aspects of academic work. The seven-plus years of working with students through the different iterations of learning activities has contributed to fostering a new generation of transformative thinkers, groups with skill sets that extend far beyond any competencies developed by students merely 10 years earlier. Combined with the LUMES knowledge to action course, there are more opportunities to understand and build capabilities as transformative sustainability scientists. However, more opportunities are still needed throughout program to grow fully engaged competent change agents.

New framing approaches will appear in coming years that even better encapsulate the complexity of socio-ecological systems. This will create the need for how to integrate them into the sustainability course. In addition, development of pedagogical approaches is not stagnant. New insights into this area will also mean new techniques to foster improved student comprehension and competency development. Finally, as societal needs change, so will the key competencies that must be nurtured in sustainability education programs at all levels. They will move beyond the key priorities of today and focus on proficiency development in areas that we still have yet to imagine. Like today, they will also present both new opportunities and challenges to grow future generations of sustainability scientists.

3.6   Conclusion

The aim of this chapter has been to present four approaches for framing complex sustainability challenges. It was done from the perspective of how the approaches are learned by graduate students in one course of an international, interdisciplinary graduate education program. Student reflections on approach competency development show that there have been challenges in achieving adequate depth in understanding of each approach. Experiences also revealed that there will be more modest ongoing challenges in student comprehension based on individual learning style preferences of students. Positive attributes for learning the framing approaches have been mainly the single topic/theme used throughout the course.

References

Avelino F, Rotmans J (2009) Power in transition: an interdisciplinary framework to study power in relation to structural change. Eur J Soc Theory 12(4):543–569

Baccarne B, Logghe S, Schuurman D, De Marez L (2016) Governing quintuple helix innovation: urban living labs and socio-ecological entrepreneurship. Technol Innov Manag Rev 6(3):22–30

Bidone ED, Lacerda LD (2004) The use of DPSIR framework to evaluate sustainability in coastal areas. Case study: Guanabara Bay basin, Rio de Janeiro, Brazil. Reg Environ Chang 4(1):5–16

Bowen RE, Riley C (2003) Socio-economic indicators and integrated coastal management. Ocean Coast Manag 46:299–312

Buhr K, Federley M, Karlsson A (2016) Urban living labs for sustainability in suburbs in need of modernization and social uplift. Technol Innov Manag Rev 6(1):27–34

Bulkeley H, Betsill MM (2013) Revisiting the urban politics of climate change. Environ Polit 22(1):136–154

Burns HL (2015) Transformative sustainability pedagogy: learning from ecological systems and indigenous wisdom. J Transform Educ 13(3):259–276

Cash DW, Moser SC (2000) Linking global and local scales: designing dynamic assessment and management processes. Glob Environ Chang 10:109–120

Cash DW, Clark WC, Alcock F, Dickson NM, Eckley N, Guston DH, Jäger J, Mitchell RB (2003) Knowledge systems for sustainable development. Proc Natl Acad Sci 100(14):8086–8091

Clark WC, Tomich T, van Noordwijk M, Guston D, Catacutan D, Dickson NM, McNie E (2016) Boundary work for sustainable development: natural resource management at the consultative group on international agricultural research (CGIAR). Proc Natl Acad Sci U S A 113(17):4615–4622

Crnogaj K, Miroslav R, Bradac HB, Omerzel GD (2014) Building a model of researching the sustainable entrepreneurship in the tourism sector. Kybernetes 43(3/4):377–393

Daniell KA, Coombes PJ, White I (2014) Politics of innovation in multi-level water governance systems. J Hydrol 519:2415–2435

Di Lucia L, Ericsson K (2014) Low-carbon district heating in Sweden—examining a successful energy transition. Energy Res Soc Sci 4:10–20

EEA (1999) Environmental indicators: typology and overview, technical report no. 25/1999, EE Agency, Editor

Evans J, Karvonen A (2014) 'Give me a laboratory and I will lower your carbon footprint!'– urban laboratories and the governance of low-carbon futures. Int J Urban Reg Res 38(2):413–430

Gari SR, Newton A, Icely JD (2015) Review: a review of the application and evolution of the DPSIR framework with an emphasis on coastal social-ecological systems. Ocean Coast Manag 103:63–77

Geels FW (2004) From sectoral systems of innovation to socio-technical systems. Insights about dynamics and change from sociology and institutional theory. Res Policy 33:897–920

Geels FW (2011) The multi-level perspective on sustainability transitions: responses to seven criticisms. Environ Innov Soc Trans 1(1):24–40

Gibson CC, Ostrom E, Ahn TK (2000) ANALYSIS—the concept of scale and the human dimensions of global change: a survey. Ecol Econ 32:217–239

Giupponi C (2007) Decision support systems for implementing the European water framework directive: the MULINO approach. Environ Model Softw 22:248–258

Hägerstrand T (2001) A look at the political geography of environmental management. In: Buttimer A (ed) Sustainable landscapes and lifeways: scale and appropriateness. Cork University Press, Cork, pp 35–58

Hruska TV, Huntsinger L, Oviedo JL (2014) An accidental resource: the social ecological system framework applied to small wetlands in Sierran foothill oak woodlands. In: The 7th California Oak symposium: managing oak woodlands in a dynamic world, General technical report PSW-GTR-251: 231–238. Visalia, California

Jerneck A, Olsson L, Ness B, Anderberg S, Baier M, Clark E, Hickler T, Hornborg A, Kronsell A, Lövbrand E, Persson J (2011) Structuring sustainability science. Sustain Sci 6(1):69–82

Kates RW, Clark WC, Corell R, Hall JM, Jaeger CC, Lowe I, McCarthy JJ, Schellnhuber HJ, Bolin NB, Dickson NM, Faucheux S, Gallopin GC, Grübler A, Huntley B, Jäger J, Jodha NS, Kasperson RE, Mabogunje A, Matson P, Mooney H, Berrien Moore B III, Timothy O'Riordan T, Svedin U (2001) Sustainability science. Science 292:641–642

Kelble CR, Loomis DK, Lovelace S, Nuttle WK, Ortner PB, Fletcher P, Cook GS, Lorenz JJ, Boyer JN (2013) The EBM-DPSER conceptual model: integrating ecosystem services into the DPSIR framework. PLoS One 8(8):e70766–e70766. http://journals.plos.org/plosone/article?id=10.1371/journal.pone.0070766. Accessed 9 Nov 2017

Klijn JA (2014) Driving forces behind landscape transformation in Europe, from a conceptual approach to policy options. In: Jongman RHG (ed) The new dimensions of the European landscape, vol 4. Dordrecht, Springer, pp 201–219

Leslie HM, Basurto X, Nenadovic M, Sievanen L, Cavanaugh KC, Cota-Nieto JJ, Erisman BE, Finkbeiner E, Hinojosa-Arango G, Moreno-Báez M, Nagavarapu S, Reddy SM, Sánchez-Rodríguez A, Siegel K, Ulibarria-Valenzuela JJ, Weaver AH, Aburto-Oropeza O (2015) Operationalizing the social-ecological systems framework to assess sustainability. Proc Natl Acad Sci U S A 112(19):5979–5984

Marshall GR (2015) A social-ecological systems framework for food systems research: accommodating transformation systems and their products. Int J Commons 9(2):881–908

McGinnis MD, Ostrom E (2014) Social-ecological system framework: initial changes and continuing challenges. Ecol Soc 19(2):374–386

Molinos-Senante M, Gómez T, Garrido-Baserba M, Caballero R, Sala-Garrido R (2014) Assessing the sustainability of small wastewater treatment systems: a composite indicator approach. Sci Total Environ 497–498:607–617

Ness B (2013) Editorial—sustainability science: progress made and directions forward. C Sustain 1(1):27–28

Ness B, Anderberg S, Olsson L (2010) Structuring problems in sustainability science: the multi-level DPSIR framework. Geoforum 41:479–488

Nevens F, Frantzeskaki N, Gorissen L, Loorbach D (2013) Urban transition labs: co-creating transformative action for sustainable cities. J Clean Prod 50:111–122

O'Higgins T, Farmer A, Daskalov G, Knudesn S, Mee L (2014) Achieving good environmental status in the Black Sea: scale mismatches in environmental management. Ecol Soc 19(3):Art 54

OECD (1994) OECD core set of indicators for environmental performance reviews: a synthesis report, organisation for economic co-operation and development: environmental monographs, Paris

Ostrom E (1990) Governing the commons: the evolution of institutions for collective action. In: Ansolabhere S, Frieden J (eds) The political economy of institutions and decisions series. Cambridge University Press, Cambridge

Ostrom E (2007) A diagnostic approach for going beyond panaceas. Proc Natl Acad Sci 104(39):15181–15187

Ostrom E (2009) A general framework for analyzing sustainability of social-ecological systems. Am Assoc Adv Sci 325:419–422

Ostrom E (2011) Background on the institutional analysis and development framework. Policy Stud J 39(1):7–27

Ostrom E, Cox M (2010) Moving beyond panaceas: a multi-tiered diagnostic approach for social-ecological analysis. Environ Conserv 37(4):451–463

Partelow S, Boda C (2015) A modified diagnostic social-ecological system framework for lobster fisheries: case implementation and sustainability assessment in Southern California. Ocean Coast Manag 114:204–217

Reis S, Morris C, Fleming LE, Beck S, Taylor T, White M, Depledge MH, Steinle S, Sabel CE, Cowie H, Hurley F, Dick JMP, Smith RI, Austen M (2015) Review paper—integrating health and environmental impact analysis. Public Health 129(10):1383–1389

Rotmans J, Kemp R, van Asselt M (2001) More evolution than revolution: transition management in public policy. Foresight 3(1):15–31

SAPECS (2016) Ecosystem services in the anthropocene: anticipating and managing regime shifts.   http://www.sapecs.org/associated-projects/ecosystem-services-in-the-anthropocene-anticipating-and-managing-regime-shifts/. Accessed 16 Nov 2016

Schneidewind U, Singer-Brodowski M, Augenstein K (2016) Transformative science for sustainability transitions. In: Brauch HG, Oswald Spring U, Grin J, Scheffran J (eds) Handbook on sustainability transition and sustainable peace, vol 10. Cham, Springer, pp 123–135

Spangenberg JH (2011) Sustainability science: a review, an analysis and some empirical lessons. Environ Conserv 38(3):275–287

Wiek A, Kay B (2015) Learning while transforming: solution-oriented learning for urban sustainability in Phoenix, Arizona. Curr Opin Environ Sustain 16:29–36

Wiek A, Withycombe L, Redman CL (2011) Key competencies in sustainability: a reference framework for academic program development. Sustain Sci 6:203–218

Wiek A, Ness B, Schweizer-Ries P, Brand FS, Farioli F (2012) From complex systems analysis to transformational change: a comparative appraisal of sustainability science projects. Sustain Sci 7(Suppl 1):5–24

Part II
Practical Approaches to Sustainability Issues

Chapter 4
The Value of Grey

Makoto Yokohari, Akito Murayama, and Toru Terada

Abstract Modern urban planning, initiated in Western Europe and North America at the dawn of the twentieth century, framed the concept of "city" as an area where no agricultural land uses should be included. In Japan, however, the demarcation between the city and countryside was ambiguously "grey" in comparison to that of Western cities. This ambiguous mixture of urban and rural land uses characterized both the fringe and the interior of Japanese cities as well. Edo, the former name of Tokyo, was already the largest city in the world in the eighteenth century with more than one million people; but at the same time, welcomed and was quite compatible with a vast amount of agricultural land that covered more than 40% of the city.

Detesting an ambiguous "grey" mixture and adoring homogeneity and clear "black-and-white" separation of land were the precepts of modern urban planning; that is, how modern urban planners framed the problem of building sustainable cities. According to such an urban planning concept, the Japanese mixed land use has long been regarded as a premodern and deniable use of land. One key feature of the 1939 Comprehensive Parks and Open Space Plan of Tokyo was developing a greenbelt surrounding Tokyo to clearly differentiate the central core of the city with its urban land uses from the surrounding countryside with its rural land uses. The City Planning Act in 1968 also aimed at achieving a clear separation of urban and rural land uses by designating Urbanization Promotion Areas (UPA) and Urbanization Control Areas (UCA) in each local municipality.

M. Yokohari (✉)
Department of Urban Engineering, Graduate School of Engineering,
The University of Tokyo, Tokyo, Japan
e-mail: myoko@k.u-tokyo.ac.jp

A. Murayama
Urban Land Use Planning Unit, Department of Urban Engineering, School of Engineering,
The University of Tokyo, Tokyo, Japan
e-mail: murayama@up.t.u-tokyo.ac.jp

T. Terada
Graduate School of Frontier Sciences, The University of Tokyo, Tokyo, Japan
e-mail: terada@k.u-tokyo.ac.jp

© The Author(s) 2020
T. Mino, S. Kudo (eds.), *Framing in Sustainability Science*,
Science for Sustainable Societies, https://doi.org/10.1007/978-981-13-9061-6_4

Cities are regarded as an entity that never creates but merely absorbs natural resources, especially food. The threat of natural disasters in Western European and North American cities is extremely low in comparison to Asian cities, and thus systems to transport food can be expected to operate with virtually little or no disruption. Cities in Asia, including those in Japan, are not afforded this luxury. They frequently suffer from sudden disruptions in transportation infrastructure caused by earthquakes, tropical hurricanes, and other natural disasters that are part of everyday life. Such a situation should therefore motivate Asian cities to maintain a redundant food supply system that can supply food even in emergencies, when logistics are disrupted for an inordinate period of time, by planning for both internal and external food supplies. Agricultural land in the city – the land likely perceived as an ambiguous "grey" mixture from the non-Asian perspective – should therefore be regarded as a reasonable and prudent land use rooted in the Asian environment. Agricultural lands also provide ecological services and are thus a crucial element for creating a sustainable city.

One conventional framing of modern civilization is its "digital approach", which tries to deductively identify fundamental elements in a "black or white" manner and then inductively synthesize such elements to re-build the entity. From such a two-value approach, the multi-value approach of "grey" has been regarded as an incomplete stage that should further be analytically identified as an entity composed of black or white elements. However, the land use mixture identified in Asian cities conveys the need for a new framing that restores and nurtures the value of grey, especially when planning for the sustainable future of the city and its surrounding region by respecting their vernacular landscapes.

Keywords Urban and rural land uses · Redundancy · Natural disaster · Urban agriculture · Food system · Resilience

4.1 Introduction

Basic theories of modern urban planning were initiated at the dawn of the twentieth century in Western Europe, where almost no threat of natural disasters as earthquakes, tsunami and tropical hurricanes was identified. Cities in the world, including those in Asia, have been taking such theories as the standard and developing themselves according the theories. However, are the theories initiated in disaster-free Western Europe cities applicable for Asian cities frequently suffering from natural disasters? Shouldn't there be alternative planning theories suitable for Asian cities? As natural disasters in European and North American cities are also increasing due to the global climate change, are planning theories initiated in Asia not suitable for their sustainable future as well? This chapter discusses an alternative framing for sustainable urban planning from one Asian perspective.

4.2   Layer Model

4.2.1   Dichotomy Versus Grey

Dichotomy is probably one of easiest approaches to understand and plan compli-
cated issues. When faced with complexity, people usually try to understand the
issue by locating it in a very simple dichotomous structure: yes or no, black or
white, right or wrong, ad infinitum. Such a dichotomous concept has been applied
to urban and regional planning. Cities in medieval Europe, often surrounded by a
wall and moat, had a clear boundary between its dense urban fabric with virtually
no green, and its surrounding wide-open rural landscapes filled with diverse types
of greenery (Fig. 4.1).

   Rooted in such a legacy, one key concept of modern urban and regional planning
initiated in Western Europe at the dawn of the twentieth century was to differentiate
urban fabric from surrounding rural areas to ensure efficiency both in urban developments in the city and agricultural production in the rural areas. At the end of the
nineteenth century Ebenezer Howard (1850–1928), an English urban planner, proposed the concept of Garden City (Fig. 4.2), a city in which people live harmoniously
together with nature. In his concept Howard stated that town and country should be
married and become a couple together, but he never meant that the two should be
mixed.

   Even though Howard said that the town and countryside should be planned
together, a distinct boundary between the two remained intact in his concept. Then

Fig. 4.1 Paris, France in sixteenth century

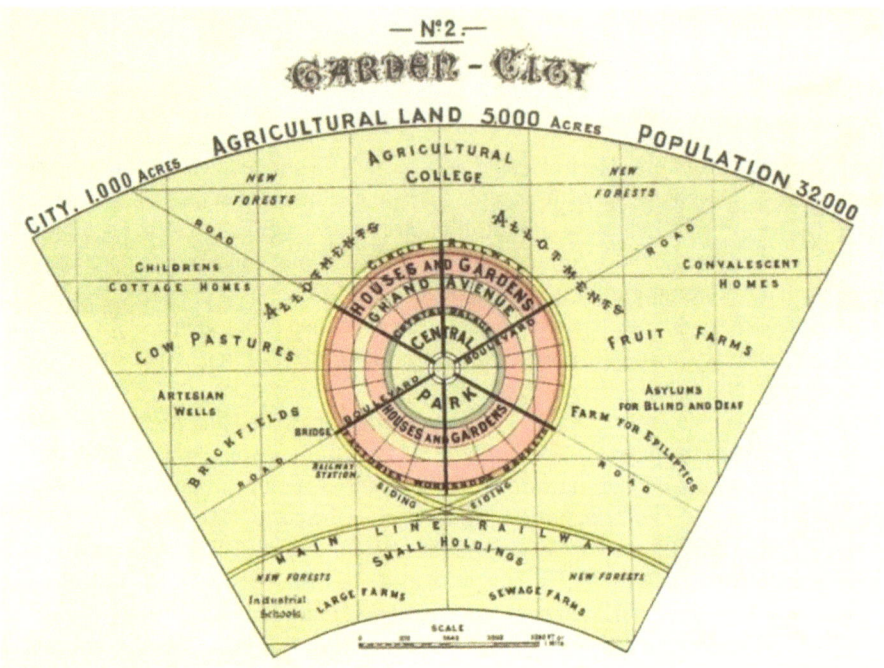

Fig. 4.2 Garden City proposed by E. Howard

came Sir Patrick Abercrombie (1879–1957), an English urban planner in charge of the Greater London Plan in 1944. In the plan Sir Abercrombie installed a greenbelt, surrounding London to curb urban expansion and clearly differentiate the urban fabric from the surrounding rural areas (Fig. 4.3). Dichotomous land use patterns came to be the international standard for modern urban and regional planning in the West. Today, many regions in the world and their cities are following the same planning system based on this dichotomous land use concept.

What can commonly be found on the fringe of Japanese cities, on the contrary, is a small-scale mixture of urban and rural land uses, which we define as "grey" landscape (Fig. 4.4). From the perspective that prefers dichotomous solutions, "grey" is often regarded as ambiguity and/or disorder. "Grey" indeed has been synonymous with uncontrolled, uncivilized, and thus undesirable solutions.

However, although a dichotomous approach provides a simple and clear but rather static and even persistent solution, "grey" allows for various shades of lightness between the extremes of black and white. If a planning concept is based on a "grey" approach, the result becomes flexible to a given condition, which leads to adaptable solutions that successfully provide "resilience" to cities and regions. The growing concern regarding natural disasters as a result of global climate change has forced cities and regions around the world to seek a new planning concept that pro-

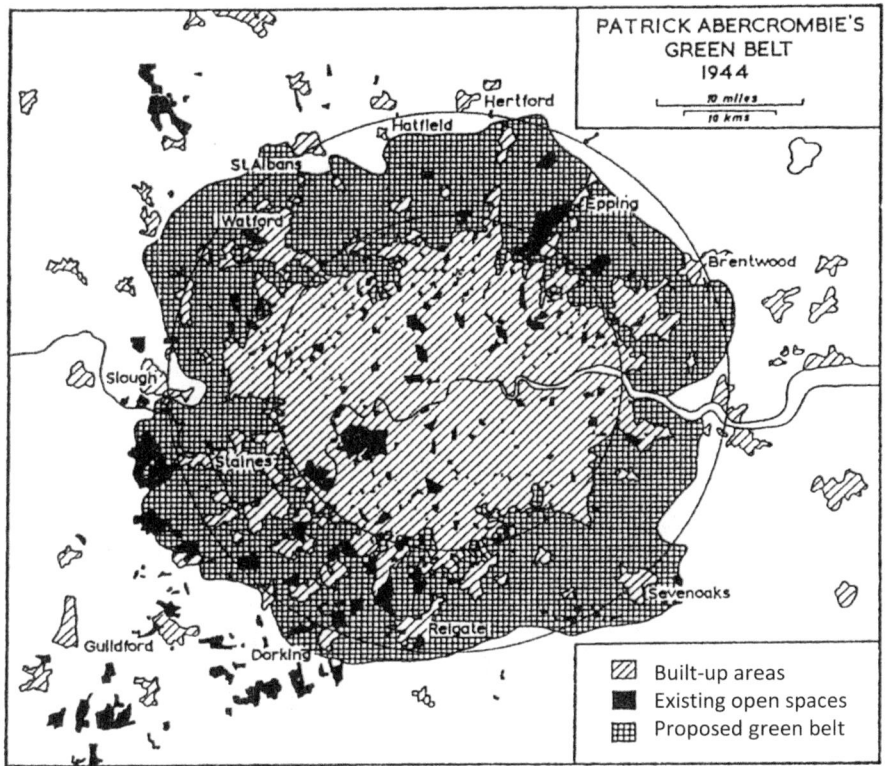

Fig. 4.3 London green belt by P. Abercombie

vides resilient solutions in responding to unanticipated catastrophes which could very well directly affect them soon. The "grey" approach is one practical answer to such demand.

4.2.2 *Landscape Patterns in Three City Regions*

To clarify the differences in landscape patterns of city regions in the West and East, we examine three major cities and their suburbs: New York City, Paris, and Tokyo. Some 15 km northwest from the center of New York City, Central Park on Manhattan Island, is a place called East Rutherford, New Jersey (population: 10,000; 10 km²). What you find in this quaint town is a typical American suburban landscape mostly comprised of *detached houses*, free-standing structures one or two stories high surrounded by wide open lawn (Fig. 4.5). Some 15 km northwest from the center of Paris, Cite Island, brings you to Argenteuil, Ille de France (population: 100,000+; 17 km²).

Fig. 4.4 "Grey" landscape in the fringe of a Japanese City

Fig. 4.5 East Rutherford, NJ, 15 km NW of New York City (Source: Google Earth)

Although the design and size of houses differ from those in East Rutherford, a similar suburban landscape with detached houses awaits (Fig. 4.6). Concerning Tokyo, however, the landscape differs somewhat. Some 15 km northwest from the city center, the Imperial Palace, lies Nerima Ward (population: 100,000+; 48 km²), which is still a part of the core area of Tokyo called *23 Wards*. Nerima

Fig. 4.6 Argenteuil, Ille de France, 15 km NW of Paris (Source: Google Earth)

Fig. 4.7 Nerima Ward, 15 km NW of Tokyo (Source: Google Earth)

is a typical residential neighborhood in the suburb of Tokyo, but includes small parcels of farmland in addition to houses (Fig. 4.7). Nerima is still in Tokyo, one of largest cities in the world that accommodates and home to more than 10 million people. Even so, within its boundary farmland parcels remain a trait of Tokyo's dense urban fabric.

Travelling 40 km northwest from New York City lies Pyramid Mountain, NJ. In addition to small villages, what is mostly found in this area is forest (Fig. 4.8). Some 40 km northwest from Paris is a village called Vigny, an area which is mostly farmland (Fig. 4.9). As for Tokyo, 40 km northwest of center city brings you to a city called Kawagoe, where you find a landscape virtually the same as that of Nerima: a landscape characterized by a small-scale mixture of urban and rural land uses (Fig. 4.10).

In New York City, representing North American cities, and in Paris, representing Western European cities, a distinct boundary between urban land use and rural land use is fixed somewhere in between 15 and 40 km from the city center. In Tokyo, which represents Japanese cities, though, no such distinct boundary

Fig. 4.8 Pyramid Mountain, NJ (US), 40 km NW of New York City (Source: Google Earth)

Fig. 4.9 Vigny (France), 40 km NW of Paris (Source: Google Earth)

Fig. 4.10 Kawagoe City (Japan), 40 km NW of Tokyo (Source: Google Earth)

between urban and rural land uses can be identified because a small-scale mixture of urban and rural land uses continues the entire distance from 15 to 40 km, and even beyond.

4.2.3   Legacy of Mixture

Edo, formerly Tokyo, is known as a city which used to be the largest in the world, accommodating over one million people at the beginning of the eighteenth century. The population density of the city was nearly five times higher than that of Tokyo today, even though houses were mostly one or two stories high. However, despite having such a massive and dense urban fabric, more than 40% of the land inside the administrative boundary of Edo was designated for agricultural uses (Fig. 4.11). Moreover, such farmland parcels were integrated into the urban fabric, not merely surrounding the city as is common in western urban design. Though an administrative boundary has existed, no physical boundary which visually separates the urban fabrics from surrounding rural land uses could be identified on the fringe of the city.

Such a legacy still continues. Today, even in the core area of Tokyo which is comprised of 23 Wards, 11 wards still maintain farmland parcels in their territory.

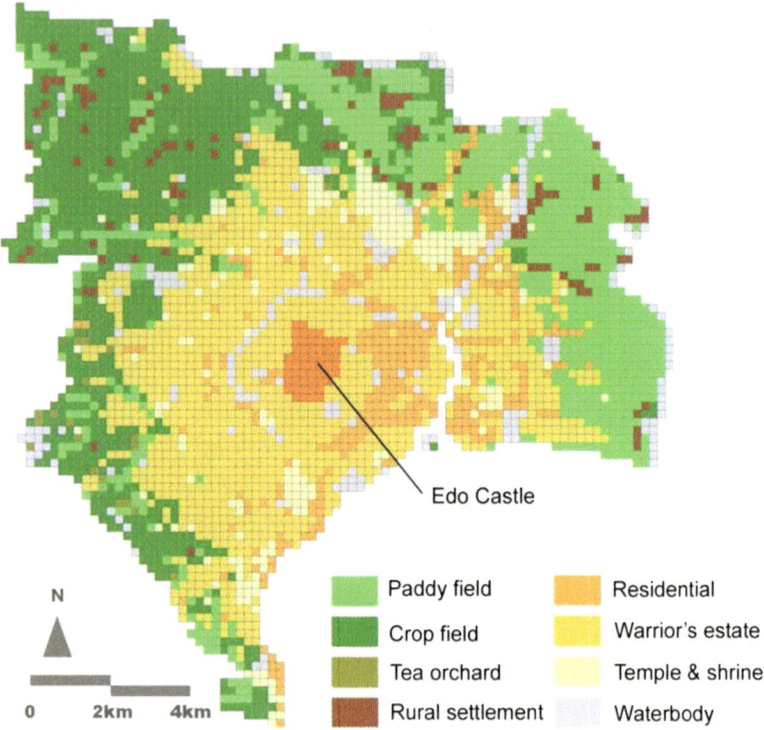

Fig. 4.11 Land use of Edo, formerly Tokyo, in early nineteenth century (Source: Fujii et al. 2002)

The amount of farmland parcels is limited: only around 3.5% of all Tokyo and 1% of the 23 Wards core area. However, although the amount is limited and the size is very small – sometimes as small as 500 m^2, smaller than a 50 m swimming pool – these farmland parcels are mostly active farmland still owned and maintained by professional farmers, not farming area for urban hobby farmers or retirees (Fig. 4.12).

4.2.4   Layer Model

What land use models are behind these realities? The Western land use model starts with drawing a clear boundary between urban and rural zones, and then cuts the land into units with homogeneous land uses. The model can therefore be character-ized as a system which provides ordered and well-controlled land uses. Japanese planners once applied this rationale to Japanese cities including Tokyo. In 1939 Comprehensive Parks and Open Space Plan of Tokyo was proposed, and one key feature of the plan was a greenbelt surrounding Tokyo to stop *urban sprawl* – the rapid expansion of the geographic extent of cities and towns – and thus realize a distinct separation of urban and rural land uses (Fig. 4.13).

Fig. 4.12   Farmland remaining in Tokyo

However, installing a greenbelt did not prove to be a success. Even if you were to look at Tokyo today from a satellite, not even a one remnant of the belt can be found. What is visible is a large-scale maze of urban fabric continuously sprawling all the way towards the mountain ranges surrounding Tokyo.

Other Japanese cities including Osaka and Nagoya also tried to install a greenbelt but they all failed because of the lack of efficient policies on the land use. Instead of a greenbelt, cities in Japan changed their policy to draw a boundary line surrounding each local municipality and not around the entire metropolitan area. The Urban Planning Act, revised in 1968, was designed to achieve such a separation. According to this Act, each local municipality was required to designate land as either one of two types: Urbanization Promotion Area (UPA), or Urbanization Control Area (UCA). UPA is the area for urban developments; UCA is, in principle, primarily for agricultural uses without conventional urban development.

But once again, distinct separation failed to be achieved. What actually happened was an incomplete separation even though a line to designate UPA and UCA was drawn around the city. Why did such a failure occur? We would argue that this situation occurred because of the layer model which the Japanese planning system had been maintaining, and not because of an inadequate application of the City Planning Act of 1968.

In short, two major layers characterize the model. First is, of course, the "Urban" layer, based on the City Planning Act of 1968, but this is not the only layer. The

Fig. 4.13 Green belt planned in the 1939 Tokyo Parks and Open Space Plan

second layer which defines the land use in Japan's urban fringe is a "Rural" layer based on the Agricultural Land Act of 1952. The Japanese agricultural system had long been based on a landlord-tenant farmer system, which prohibited Japanese agriculture from becoming modernized and thus caused tenant farmers to endure extremely low income. The Agricultural Land Act aimed to eliminate such a system and modernize agriculture by making farmland available to all tenant farmers. The Act, however, also prohibited non-farmers from owning their own farmlands because the former landlord-tenant farmer system could very well have been revived if farmlands were bought by non-farmers, especially by enterprises, and rented out to farmers.

The Agricultural Land Act can therefore be interpreted as an act that aimed to draw a line between *people*: sharply differentiating farmers and non-farmers. The Urban Planning Act of 1968 was an act to draw a line between *land* use differentiating urban (UPA) and rural (UCA) land uses. Japanese did not ignore but have carefully been obeying the regulations. However, because these two layers followed different orders – people-oriented versus land-oriented – a chaotic-looking situation occurred when these two were overlaid. The situation should not be labelled "disordered" because each layer is well controlled albeit following different orders. Order is there, but is not visible at a glance. The layers must be separated to understand the order of each layer, which is called an underlying "hidden order" (Ashihara 1989).

4.3 Shaping the "Grey Urban Environment"

4.3.1 "Grey" in Urban Context

"Grey" in Sect. 4.1 mainly focuses on the mix of urban land uses (residential, commercial, and industrial) and rural land uses (farmland, forest, etc.). In the Sects. 4.2 and 4.3 we take a closer look into the urban area, "Grey" is interpreted more broadly: (1) diverse types of "grey", not only "urban-rural"; (2) mix of uses, forms, and densities; (3) border between private and public; and (4) flexible transformation of land uses. These represent the elements of adaptable planning embedded in the Japanese urban planning system.

4.3.2 Grey Urban Environment in Tokyo

The view of inner-city and suburban areas of Tokyo from the observatory of the Tokyo Metropolitan Government Building located in Shinjuku, one major urban center in central Tokyo, well illustrates the grey urban environment of Tokyo (Fig. 4.14). A mix of buildings – large buildings along skeletal roads and small buildings of different sizes and uses – is seen. The difference between this view and the view of European or North American cities from tall buildings is immediately noticeable.

4.3.3 Grey Urban Environment in Tokyo

The view of inner-city and suburban areas of Tokyo from the observatory of the Tokyo Metropolitan Government Building located in Shinjuku, one major urban centers in central Tokyo, well illustrates the grey urban environment of Tokyo (Fig. 4.14). A mix of buildings – large buildings along skeletal roads and small buildings of different sizes and uses – is seen. The difference between this view and the view of European or North American cities from tall buildings is immediately noticeable.

In a residential area near Ikebukuro, another major urban center in central Tokyo a little north of Shinjuku, large-scale redevelopments stand in contrast to existing small-scale urban environment (Fig. 4.15). A few minutes east from that location, an urban environment in transition is found: aging buildings and unmanaged vacant plots (Fig. 4.16). To the south of the Ikebukuro urban center, construction of a new road along existing streetcar tracks is in progress (Fig. 4.17), and is also changing surrounding land uses and buildings. The road is being constructed *after* urbanization, so many existing buildings must be demolished to construct the new road, a prime example of modern infrastructure.

Fig. 4.14 Inner-city and suburban areas of Tokyo

Fig. 4.15 Residential area near Ikebukuro urban center

Fig. 4.16 Few minutes walk from the site of Fig. 4.15

Fig. 4.17 New road construction in an existing urban area

About 10 km northwest of the Ikebukuro urban center is a unique urban landscape: a mix of farmlands, detached housing, apartments, and condominiums in an ongoing urbanizing area (Fig. 4.18). Typical urban sprawl results in the mix of land use and vulnerable infrastructure. Toward the edge of the urban area is a mix of unmanaged buildings and vacant plots caused by population decline and aging (Fig. 4.19).

But, not all urban areas in Tokyo are grey. Master-planned urban (re)developments and the installation of skeletal infrastructures are found in existing urban areas. Okata and Murayama (2011) describe Tokyo's urban form more in detail.

4.3.4 *Japanese Urban Planning System*

The Japanese urban planning system consists of four elements: (1) master plans for city planning areas and municipalities; (2) land use regulations (area division and zoning); (3) development of urban infrastructure such as roads, parks, water works, and sewage systems; and (4) urban development projects such as land readjustment and redevelopment (see MLIT (2003) for the details of the Japanese urban planning system). It should be emphasized that urbanization often progresses prior to formal urban planning and development under such a system because urbanization was rapid. The illustrations of urban planning and

Fig. 4.18 10 km NW of Ikebukuro urban center

Fig. 4.19 Fringe of Tokyo's suburban city

development in an actual city show that urban development and road development occur in small segments, resulting in a patchwork of various urban areas connected by a continuously expanding road network.

The land use planning concept is shown in Fig. 4.20. First, the City Planning Area where the City Planning Act is in effect is designated. We divide the City Planning Area into Urbanization Promotion Area (UPA) and Urbanization Control Area (UCA). UPA is divided into 12 land use zones including 3 commercial, 3 industrial, and 6 residential zones. District Plans are developed for some areas that need more detailed, special regulations and projects. Buildings are regulated through density and from regulations. We must emphasize that most zones within UPA in the Japanese land use planning system allow a mix of uses including agricultural even in Urbanization Promotion Areas. That is, the nature of the Japanese urban land use system itself includes grey or vague aspects. Murayama (2016, 2017) explains the Japanese urban land use planning system and practices more in detail.

Urban development projects in the Japanese urban planning system contribute to shaping the grey character of the Japanese urban environment. One such urban development project is called "mini" development. Residential or agricultural

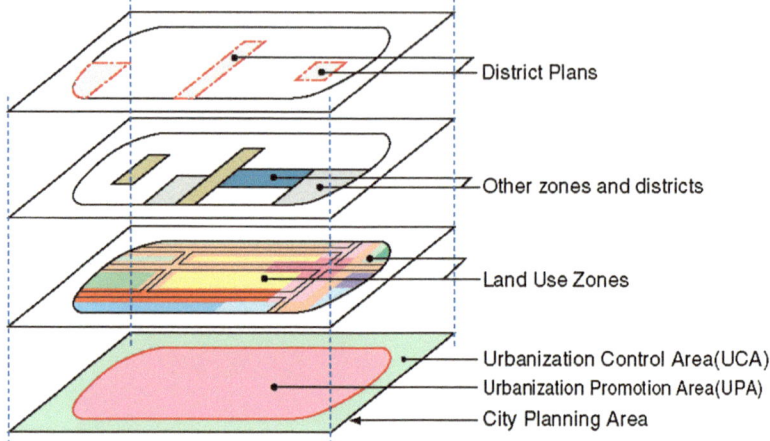

Fig. 4.20 Concept of urban land use planning (Adapted from "Introduction of Urban Land Use Planning System in Japan, City Bureau, Japanese Ministry of Land, Infrastructure and Transport (2003) 国土交通省都市局都市計画課提供)

plots of less than 1000 m² can be developed as, for example, 8 housing plots with a dead-end street generally 4 m wide or a bit wider. Urban sprawl areas such as Tagara in Nerima Ward, Tokyo, a former agricultural area, are urbanizing through "mini" developments in tandem with incremental development of streets and parks (Fig. 4.21). At the same time, some farmers continue to maintain their farmlands to produce vegetables and fruits. As urbanization advances, streets and parks are developed incrementally, thus slowly transforming an agricultural area into an urban residential area. Arterial roads – high capacity urban roads – are also constructed. In this kind of incremental development, new houses are constructed little by little over a long period of time. Home-buyers mostly in their 30s and 40s will move into the area gradually thus contributing to the diversity of resident age groups. In addition, this kind of development leaves opportunities for urban farming.

4.3.5 Uniqueness of Japanese Urban Planning

Japanese urban environment can be characterized as the islands of planned development in the sea of urban sprawl where urbanization occurred without master-planned infrastructure. The formal approach of Japanese urban planning and development has been to increase the areas of planned development through urban development projects and to install skeletal infrastructure in already-sprawled urban areas. What results is vast areas of grey urban environment.

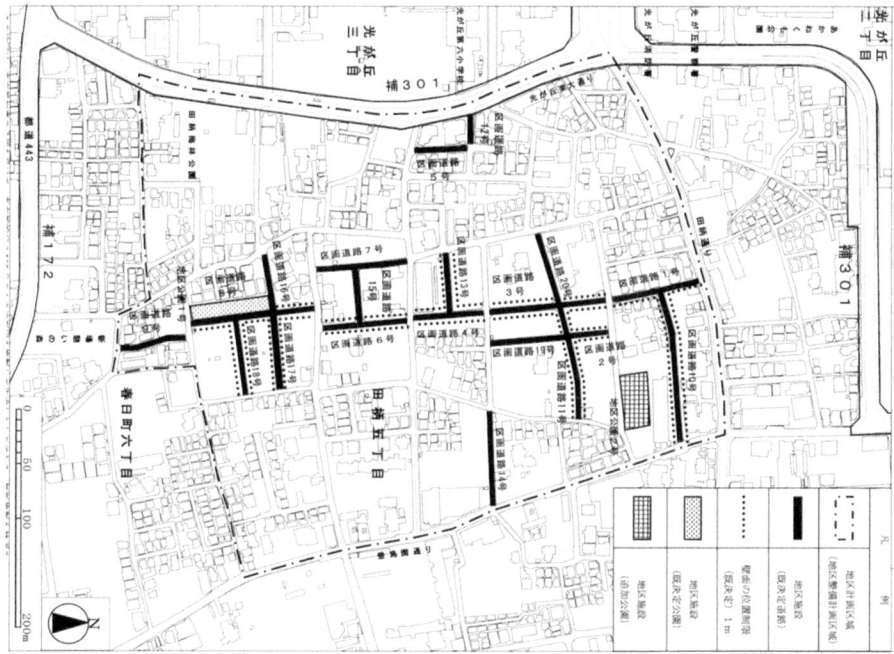

Fig. 4.21 District plan in urban sprawl area (Example in Tagara, Nerima Ward, Tokyo from Nerima Ward website)

4.4   Enhancing the Values of Grey Urban Environment

4.4.1   High Density Urban Areas in Tokyo

In the previous section, we introduced the urban sprawl area of Tagara. Returning more toward the center of Tokyo, a belt of high density urban areas where urbanization and densification occurred without master-planned infrastructure (Fig. 4.22) is found. These high-density urban areas are vulnerable because of the susceptibility of fire especially when major earthquakes occur. The Hanshin-Awaji Major Earthquake in 1995 is a excellent example. Since then, much effort has been put into improving the physical environment of these high density urban areas: widening roads, creating additional open spaces, redeveloping wooden buildings, among others.

Despite the vulnerability, this kind of high-density urban area is popular because of good access to urban centers of central Tokyo, small-scale urban environment; and the availability of affordable housing, active commercial areas, walkable neighborhoods, urban culture, etc. This vibrant commercial area in

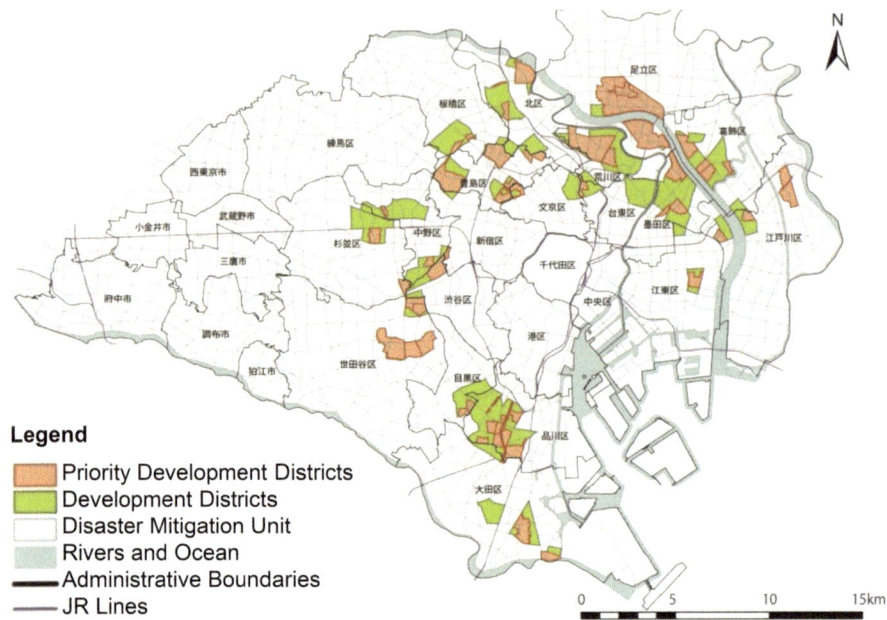

Fig. 4.22 Tokyo's 23 wards and some western suburban cities (Source: Bureau of Urban Development, Tokyo Metropolitan Government 東京都都市整備局提供)

Koenji, Suginami Ward, Tokyo is grey in a way (Fig. 4.23). The border between private and public is unclear, and merchandise, and tables and chairs are illegally placed on the street. This borderless relationship between shops, restaurants, and the street is attractive for urban dwellers. But in formal urban planning, a plan has been developed a long time ago to modernize this commercial area by constructing a new road in the existing urban area. If this road is actually developed, most of the shops and restaurants will be relocated resulting in a totally different urban environment. The modernized commercial area will have wider roads that clearly separate pedestrians and automobiles, and larger buildings. Thus, the characteristics of the vibrant, small-scale commercial area that urban dwellers enjoy now will disappear.

4.4.2 Modernization: The Only Solution?

Jane Jacobs (1906–2006), a famous North American journalist and activist who often wrote about preserving urban neighborhoods, raised this question in the 1960s. She fought against new big-money developments and emphasized that existing urban areas with higher population density, mixed uses, older buildings, and

Fig. 4.23 Vibrant and grey commercial area (Example in Suginami Ward, Tokyo)

short blocks are much more attractive than the redeveloped sites and should be protected from modern redevelopment.

In recent years, North American cities have come to recognize the value of urban farmland. This mix of residential and agricultural land is already common scenery in sprawling urban areas in Japanese cities. Many Japanese planners consider sprawled urban areas – "grey" urban environment – as a failure of modern urban planning, and try to improve or even redevelop these areas. Re-evaluating the positive aspects of this grey urban environment may very well provide alternative solutions to a sustainable and resilient city.

4.4.3   New Values and Ideas to Stay Grey

At this point we would like to introduce three cases with new values and ideas to ensure a grey urban environment. The first case is a residential area with urban farmland. Many of the sprawled suburban areas in Japanese cities are residential areas with farmlands like Nishi-Tokyo City, Tokyo (Fig. 4.24). Here the loss of urban farmlands or productive green spaces is related to the individual circumstances of aged farmers. Once a plot of urban farmland can no longer be maintained by a farmer, it will likely become a "mini" development of small-detached houses for economic benefits.

But recently, the market for detached houses seems to be declining because of the increase in construction costs and the changing attitudes toward home ownership. An alternative approach to deal with the loss of urban farmland must be found. Urban farmlands are important in maintaining the quality of sprawled urban area because the area is unequipped with sufficient streets and parks.

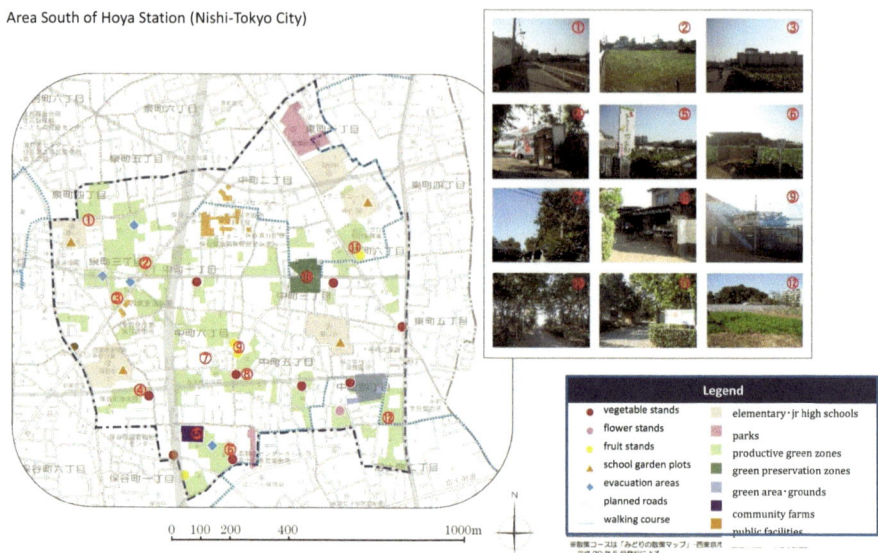

Fig. 4.24 Urban farmland in Nishi-Tokyo City, Tokyo (Source: *Model Plan for Urban Development: City and Farming Industry Co-Exist* (2010) Nishi-Tokyo City Government 「都市と農業が共生するまちづくりモデルプラン」 西東京市生活環境部産業振興課, 2010年)

Using the transfer of development rights, a technique that encourages the transfer of growth from places where a community would like to see less development to places where a community would like to see more development, is one idea for conserving productive green space (Fig. 4.25). A mix of productive open space and mini detached houses represents the present situation. If no action is taken, productive green space will be lost to mini-detached housing developments; or in the case that no market for mini detached houses exists, land will undoubtedly be abandoned. Presupposing the existence of a market for eco-collective housing for rent or sale, higher density housing could be built along newly constructed arterial roads. Arterial roads are already planned and are to be constructed to form a better road network. Through the transfer of development rights, green space inside the superblocks surrounded by arterial roads can be conserved, thus contributing to the overall quality of the urban area.

The second case is a low-density residential area with community-managed forest. Fujimaki-cho in the eastern hills of Nagoya City, Aichi Prefecture, is designated as an area for a future urban park (Fig. 4.26). But the park is unlikely to be developed because of the shortage of public funds. Nearly 200 households live in the future park area with a minimum urban infrastructure. Streets are partly unpaved and houses are not connected to the city's sewage system. The forests in Fujimaki-cho are maintained by resident volunteers. The community workshops we organized there conducted lengthy discussions about the current issues and the future scenario of Fujimaki-cho. When a city has no public funds to purchase and maintain such land for the future

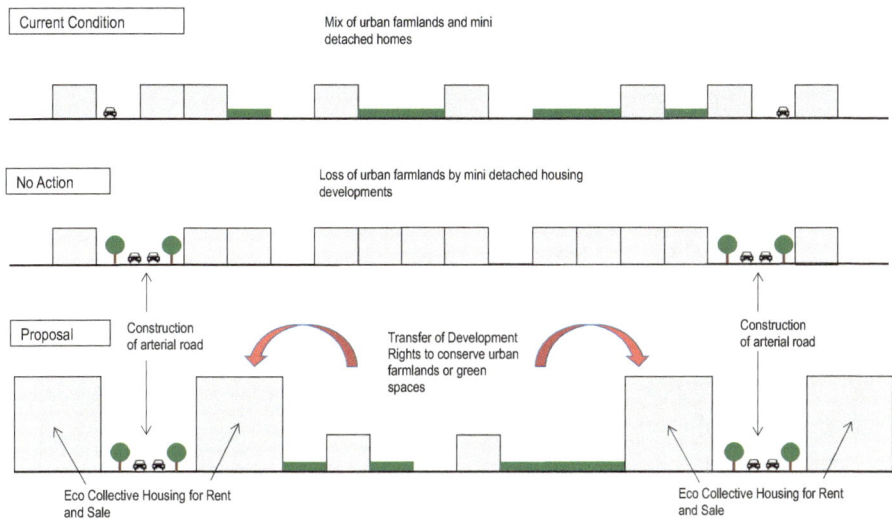

Fig. 4.25   The concept of transfer of development rights in urban sprawl area

Fig. 4.26   Low density residential area in undeveloped urban park (Fujimakicho, Nagoya City, Aichi Prefecture)

park including the plots of nearly 200 existing households, the help of residents and citizens is indispensable for any realistic solution that conserves urban forests.

A scheme is now in place that postpones urban park development in the highly inhabited part of the future park, and reduces the area of urban park development (Fig. 4.27). In the downsized area, the implementation of the city's urban park development plans can be accelerated. Before such urban park development starts

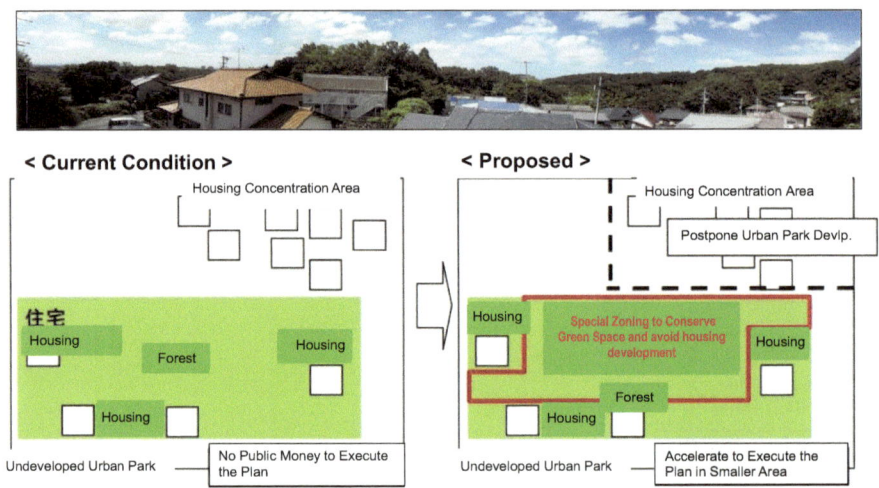

Fig. 4.27 Reorganization of urban park and residential area (Source: Greenification and Public Works Bureau, Parks Department, Parks Project Division, Nagoya City 名古屋市緑政土木局緑地部緑地事業課提供)

Fig. 4.28 Re-design of de-intensifying suburban residential area (Source: 名古屋市緑政土木局緑地部緑地事業課提供 Greenery Section, Green Policy Engineering Dept. Nagoya City)

in the forest area, however, a special zoning rule that conserves the forest and prevents new housing development should be designated.

The third case is about managing depopulating suburban residential areas (Fig. 4.28). If no action is taken, unmaintained buildings and land parcels will be generated unpredictably because of population decline and aging, thus decreasing the property value of the residential area. Through well-planned measures, creating a managed suburban residential area with lower density and higher ratio of green space is possible. Such measures will also contribute to making the entire urban form of the city more compact. The measures include, but are not.

limited to, the assembling of neighboring plots, the greening of vacant plots, the trading of plots in the chances of building reconstruction, and housing design with more open spaces.

There should be many other ideas to re-evaluate and manage a grey – actually "green" in a sense – urban environment to create sustainable urban neighborhoods. In any case, such transformation of space or physical environment should be well planned and well designed.

4.5   Shaping the "Urban-Rural Grey"

4.5.1   Land Use Transformation in Suburban Tokyo

The National Population census reported that Tokyo experienced rapid population growth from 3.7 M in 1920 to 11.4 M in 1970 mainly due to rural migration (Statistics Japan 2000). During this period, urban expansion continuously occurred in peri-urban rural areas. Through the process of urban expansion, the rural areas developed before World War 2 have already been integrated into the current urban fabric of Tokyo center. The rural areas developed after World War 2, however, have formed the current residential bed-town communities in suburban Tokyo.

Figure 4.29 shows an example of typical land use changes in suburban bed-town communities. The aerial photos cover the urban fringe of Funabashi city in Chiba prefecture some 30 km east from Tokyo's center (Imperial Palace). In 1947, Funabashi's urban fringe was rural: farmland and forest dominated. Tokyo's urban expansion had reached Funabashi between 1947 and 1970, and residential communities had developed along with construction of inter-city railway infrastructure. Up until 1997, residential development continued and this shapes the foundation of current urban fabric. Between 1997 and 2016, urban expansion moderated because of stabilization of the increasing population, but small housing developments still continue to appear. Consequently, a scene typically in Funabashi is a small-scale mixture of farmland, forest, and housing. It seems that the mixture is a result of urban sprawl on rural land without any (or ineffective) concern for land use regulation.

4.5.2   Area Division System and Agricultural Promotion Regions

However, urban-rural mixed land is basically controlled by land use regulations, from both urban and rural planning perspectives. Area Division System (ADS) is the land use regulation designed to make a boundary between urban and rural areas

Fig. 4.29 Land use changes from World War 2 to present (Location: Funabashi, Chiba, Japan). (©
Geospatial Information Anthology of Japan (1947, 1970, 1997) © 2017 Google, ZENRIN (2016))

from an urban planning perspective. In 1968 when urban sprawl was accelerating,
the City Planning Act was amended to initiate and area division system. Under this
system, a local municipality can divide an urban planning area into two areas: UPA
(Urbanization Promotion Area) and UCA (Urbanization Control Area) (Nakai
1988) (Fig. 4.30). UPA is the area where urbanization is promoted and aims to be
developed within 10 years. Once farmland or forest is included into UPA, the land
is regarded as potential land for future development.

Expectations of future development lead to drive land prices up significantly
which leads to easier conversion of farmland or forest to housing or other urban
land uses. UCA, on the contrary, is an area where urbanization is regulated and
which aims at conserving rural settings and agricultural activities. Land prices in
UCA are considerably lower compared to those of UPA because of land use regulations concerning future development. The top-right illustration of Fig. 4.30 shows
the actual implementation of area division system to the urban fringe of Funabashi.
Because the separation looks like a line drawn between UPA and UCA-commonly
called "senbiki" in Japanese, which literally means "draw a line". Even the shape
of the line is not simple: the line makes a sharp contrast between UPA and UCA in
terms of building density or farmland ratio, for example.

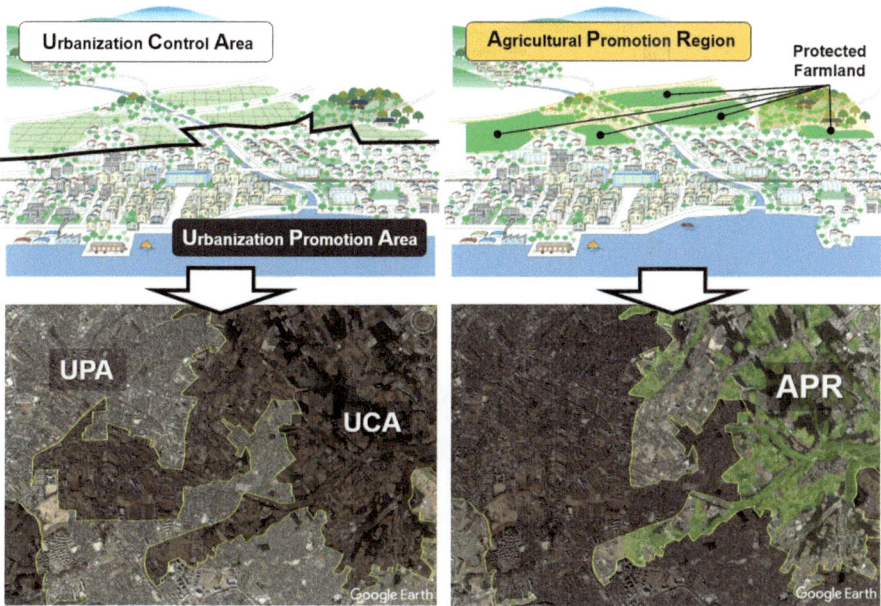

Fig. 4.30 Implementations of the land use policies (Location: Funabashi, Chiba, Japan). (© Ministry of Land, Infrastructure, Transport and (Top illustrations) © 2017 Google, ZENRIN (Bottom aerial photos))

Another dimension of land use control is rural planning. Designating an area as an Agricultural Promotion Region (APR) is the measure with most impact. APR is basically a zoning for promoting agriculture in rural areas. When applied to peri-urban areas, though, farmland protection takes on a more significant role. Once an area becomes APR-designated, the productive farmlands inside APR are protected which, in principle, may not be changed to any other land use (Fig. 4.30). Protected farmland is crucial for farmers who want to continue agriculture near cities. Rice farmers especially can conserve irrigation systems by designating protected farmland. The bottom-right illustration of Fig. 4.30 shows the implementation of APR and protected farmland in Funabashi. Comparing the ADS and APR systems reveals that these two systems are like two sides of a coin. Overlapping designation of UCA and APR is the strictest control of land use, whereas sole designation of UCA permits urban-rural mixture.

4.5.3   Productive Green Land

Although UPA is an area for urban development in theory, small farmland patches can be found in UPA and this makes a unique landscape which is a farmland-residential mix. Most such farmlands are protected as Productive Green

Fig. 4.31 Distribution and scenery of productive green lands (Location: Kashiwa, Chiba, Japan). (© City Planning Section, Kashiwa City (Top left) © 2017 Google, ZENRIN (Top right))

Land (PGL), which is the special protection for farmland in UPA based on the Productive Green Land Act established in 1974 and amended in 1991 (Fig. 4.31). Extremely high land prices and taxes of the Tokyo region makes keeping farmland in UPA nearly impossible (Yagasaki and Nakamura 2010). At the same time, however, some farmers in UPA are willing to continue their agriculture livelihood. The most significant role of the act is to reduce the tax burden on landowners (farmers). Once their farmlands are designated as PGL, land may not be sold and used for any other purpose. The farmers also must continue agriculture for at least 30 years from the designation. Most PGL were designated in 1992, the year the act was enforced (Tsubota 2006). Accordingly, there is possibility that a great number of PGL will be dissolved around 2022 because of the 30-year mark from the initial 1992 designations. This predictable issue is called the "2022 Productive green land problem" and recognized as a critical problem for farmland protection in cities (Terada 2017a).

4.5.4 Hidden Order in Planning System

Peri-urban landscape in Tokyo looks like chaotic urban-rural land use mixture, but as explained previously, the fact is that the mixture is controlled from both urban and rural planning points of view. Japanese land use control systems are strictly implemented, and overlapping of the systems permits urban-rural mixture as a case

Actual visual entity

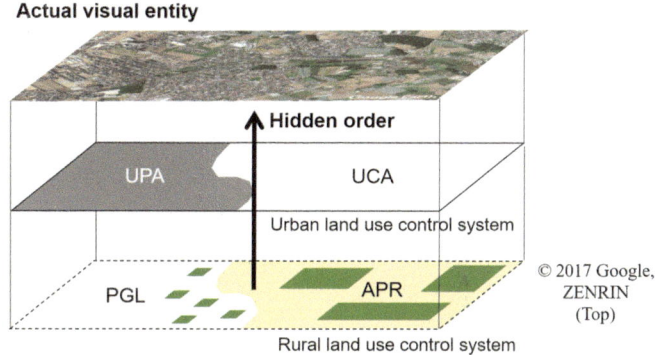

© 2017 Google,
ZENRIN
(Top)

Fig. 4.32 Land use control systems creating hidden order in actual land use pattern

of Productive green in UPA (Fig. 4.32). The variety of overlaps creates several patterns of urban-rural mixture, but each land use control system is well coordinated to be compliant. This phenomenon is called "hidden order" in landscape, an idea originally developed by Japanese architect Yoshinobu Ashihara, applying to Japanese architecture as a metaphor for culture to explain an insider's look at the apparent lack of order of Tokyo (Ashihara 1989).

4.6   Enhancing the Value of Urban-Rural Grey

4.6.1   Growing Vegetables as a Retiree Lifestyle

Japan has a rapidly aging population. Many elderly people living in the suburbs of Tokyo belong to the baby boomer generation born just after World War 2 (Fig. 4.33). They worked in Tokyo and commuted to their companies from their suburban home. Currently, the number of retirees is increasing and quite a few people have started growing vegetables in their neighborhood as a part of their retiree lifestyle. This is partly because of the proximity of their homes to farmland in urban-rural mixture of suburban Tokyo.

Those who want to start farming have several options. If they seriously intend to contribute to the agricultural industry, they can support a professional farmer as a part-time worker. Or, nowadays they can even become professional farmers with the support of local municipalities. However, the easiest option for becoming involved in farming is becoming a hobby farmer, a person who enjoys farming for leisure and as a non-profit activity. A reasonable option for a hobby farmer to start farming is to rent a small plot(s) of allotment gardens (10–30 m^2) which are currently provided by various organizations including municipalities, agricultural associations (Japan Agriculture Cooperatives), or even private industry entrepreneurs (Fig. 4.34). Most commonly, allotment hobby farmers grow vegetables by themselves, but currently an alternative includes expert guidance by a professional farmer or gardening expert from a private company.

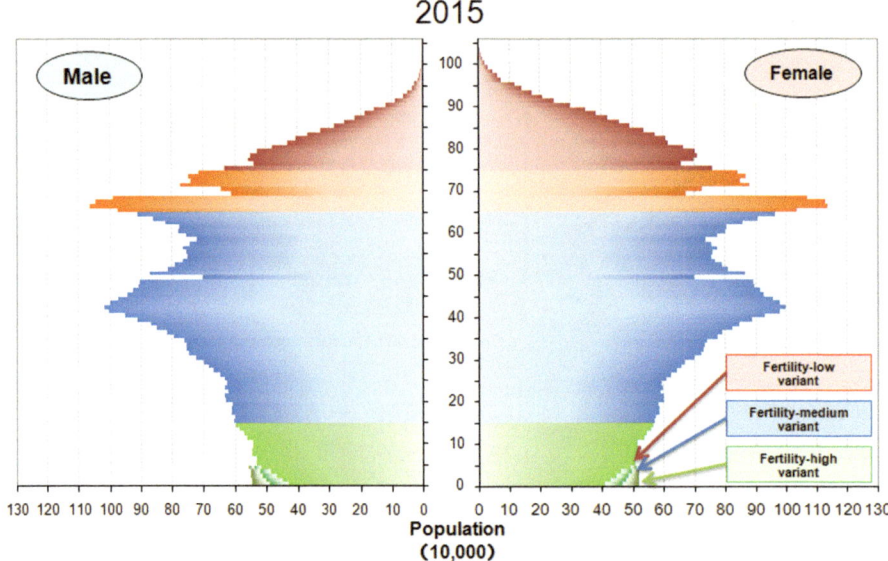

Fig. 4.33 Population pyramid for Japan in 2015. © National Institute of Population and Social Security Research. (Sources: Census (1920–2015)

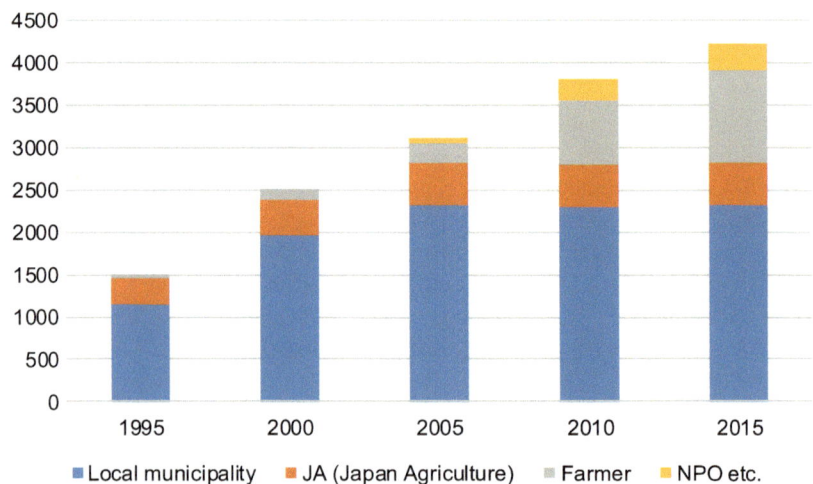

Fig. 4.34 Changes in the number of allotment gardens in Japan. (Source: Ministry of Agriculture, Forestry and Fisheries)

4.6.2   Food Provisioning from Hobby Gardens

Hobby farming in allotment gardens reaps a variety of benefits, the most direct of which is access to fresh food (Pothukuchi 2004). While quantifying food production in allotment gardens is regarded as a valuable assessment, it remains unknown (Gittleman et al. 2012). Therefore, we tried to identify actual yield from two selected allotment gardens in the Tokyo region (Tahara et al. 2011). One is the Hagidai garden in Chiba city, which is a typical allotment farm without guidance of professionals (non-guided type). The other one is the Shiraishi garden in Nerima ward, which is Japan's first allotment garden with farmer's guidance (guided type) (Fig. 4.35).

The types of vegetables planted in each garden are diverse (Fig. 4.36). Warm climate and four distinct seasons enable gardeners to grow both summer vegetables (tomatoes, eggplants, edamame, okra, corn, etc.) and winter vegetables (potatoes, daikon radish, broccoli, onions, cabbage, etc.). The allotment farmers plant a large variety of vegetables in small amounts characteristic of hobby farming. Yields were identified from direct weight measurements by gardeners randomly selected from each garden (Fig. 4.37). For the guided type, farmers carefully prepare soil before planting to be rich and homogeneous for all plots, and decide the vegetable planting pattern commonly applied to all allotments.

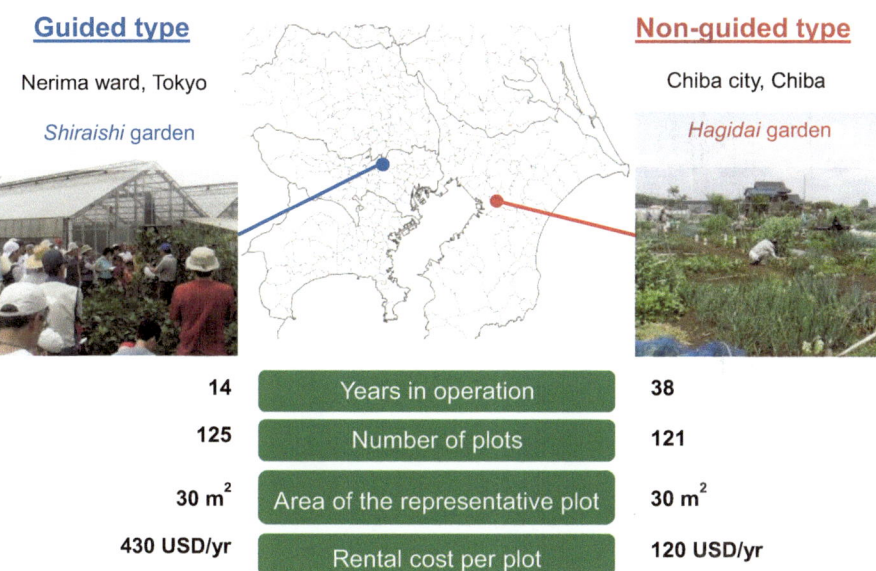

Fig. 4.35   Location and basic figures of the case study gardens (Tahara et al. 2011)

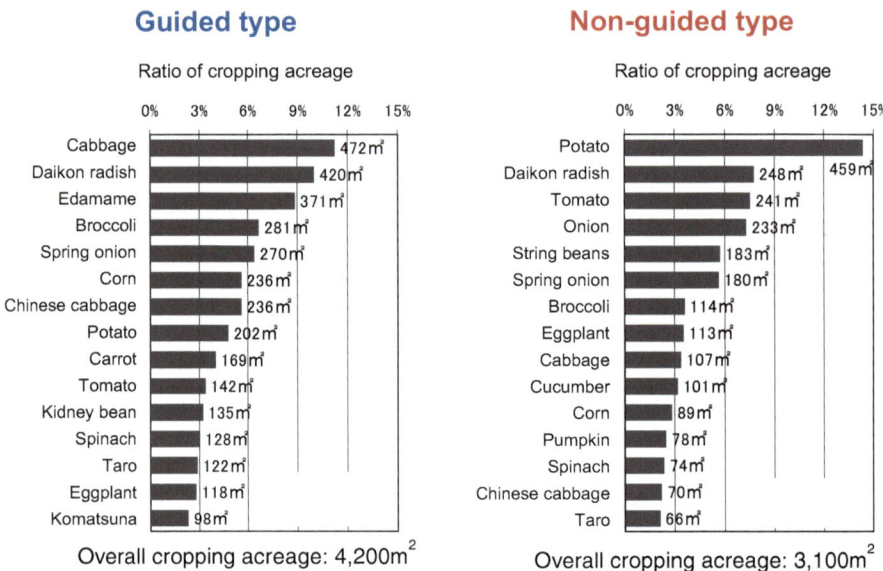

Fig. 4.36 Summary of the vegetable planting patterns of the case study gardens (Tahara et al. 2011)

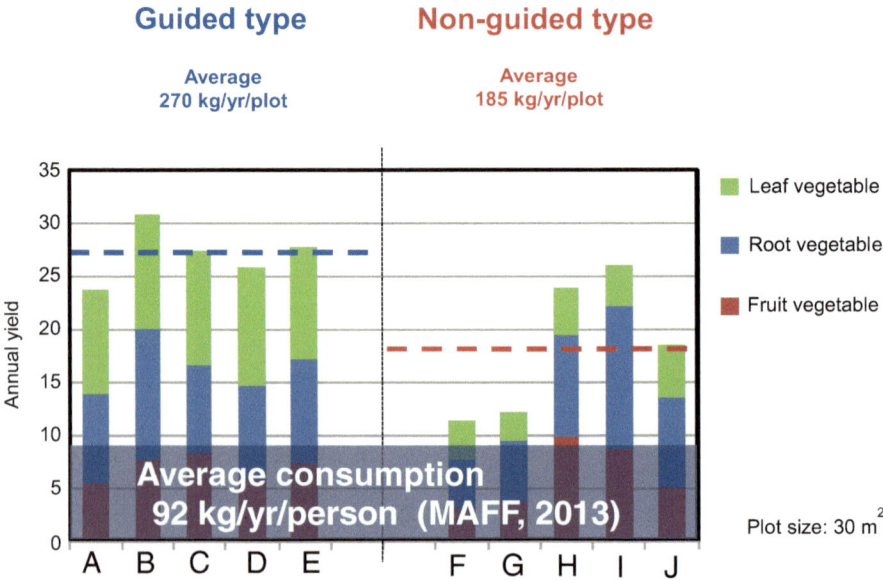

Fig. 4.37 Annual vegetable yield of the selected 10 examinees (Tahara et al. 2011)

Guided type

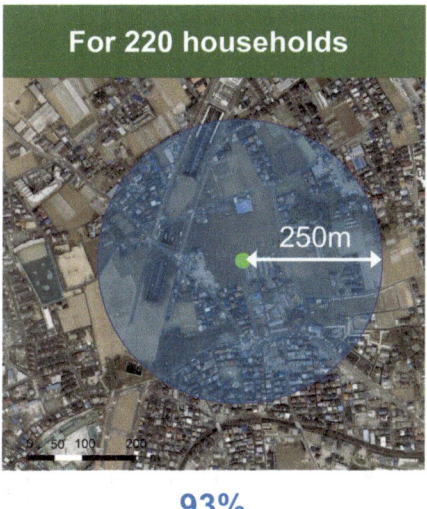

For 220 households

250m

93%

Non-guided type

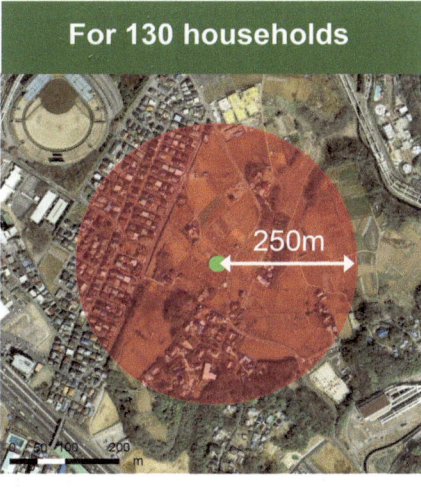

For 130 households

250m

54%

Fig. 4.38 Self-sufficiency in vegetables in a neighborhood community (Tahara et al. 2011). (Aerial photos © Geospatial Information Anthology of Japan)

Such farmer guidance improves the yield and stability of guided-type hobby farms. Compared to the current amount of average vegetable consumption per person (89 kg/year) (MAFF 2016), guided-type hobby farms can produce three times more. The non-guided type can produce only two times on average. Simply speaking, at least both garden types can produce enough vegetables for self-consumption for the hobby farmers themselves; and in most cases, they can share their excess with family or neighbors.

Japanese allotment gardens are vegetable-oriented. When considering food security in cities, allotment gardens can contribute to producing emergency food and nutrients for neighborhood community, especially when the food transportation is disrupted by natural disasters (Sioen et al. 2017). The numbers in Fig. 4.38 shows the ratio of self-sufficiency in vegetables supplied from the two example gardens to meet the demand (current vegetable consumption) of the immediately surrounding communities (Fig. 4.38). The numbers are considerably high, although the gardens are located in the densely populated Tokyo region. Allotment gardens may not just be a substitution for urban greenery or urban open spaces, but be part of the productive landscape (Viljoen and Howe 2012) that links urban farming and the local food system.

Fig. 4.39 Typical scenery of abandoned Satoyama

4.6.3   Satoyama Woodland as Community Biomass Energy Source

Forest is also one agricultural land use. Tokyo suburbs used to be rural areas, thus existing forests in Tokyo's present-day suburbs were historically maintained by the agricultural community. Such forest is called *satoyama* woodland. Satoyama is a word coined by combining village (*sato*) and mountain (*yama*). Satoyama woodland is the woodland that rural communities historically maintained for harvesting fuelwood or other organic materials to sustain their livelihood. Traditional management practices create habitats for diverse flora and fauna that can survive only under human disturbance. Satoyama woodland is a biodiversity-rich semi-natural ecosystem that benefits both human and nature (Takeuchi et al. 2012).

The widespread use of fossil fuels in today's world has caused many satoyama woodlands to lose their role of producing biomass fuels. Most satoyama woodlands today are abandoned because of the loss of economic value. This leads to changes in the ecosystem that had been maintained by human disturbance. Declining biodiversity in satoyama woodland is regarded as one crisis in the National Biodiversity Strategy (MOE 2010), and this abandonment causes social problems including illegal dumping (Fig. 4.39).

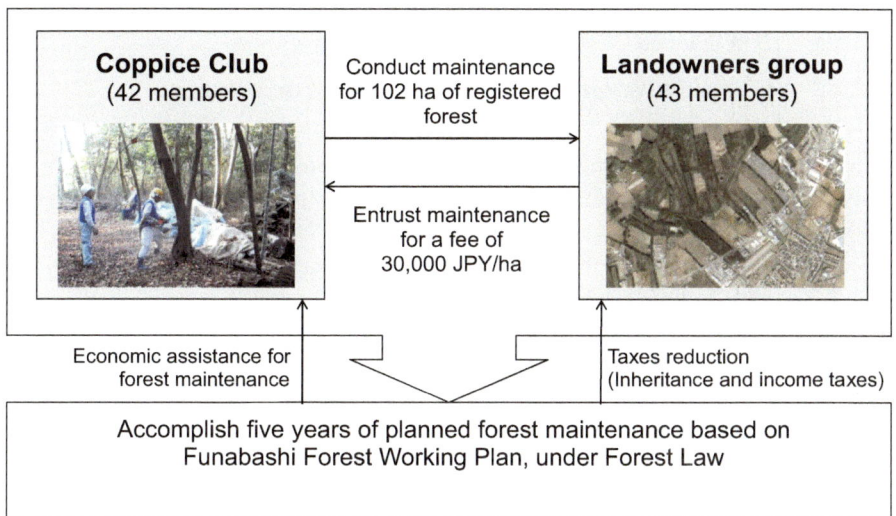

Fig. 4.40 Schematic overview of Funabashi Forest working plan (Terada 2017b)

Many citizen groups endeavor to address this issue of re-maintaining satoyama woodland as an urban productive landscape. Coppice Club, satoyama-friendly group organized by retirees in Funabashi, Chiba is a prime example of such citizen groups. They are attempting to restore the maintenance of an abandoned satoyama woodland of some 100 ha based on a contract with landowners (Fig. 4.40) (Terada 2017b). As a result of active maintenance, these retirees are producing a large amount of biomass. If their maintenance techniques were applied to the entire forest in Funabashi city (720 ha), around 1000 tons of biomass are estimated to be harvestable, contributing to a 10% self-sufficiency in heat energy in the neighborhood communities in urban-rural mixed areas (Matsumoto et al. 2011).

The biggest barrier in making biomass utilization feasible is the high cost of biomass transportation. Proximity of satoyama woodland and urban areas may tackle this barrier by minimizing the distance between satoyama woodland and biomass heat or electricity plants in urban areas. Related to this, it is estimated that the biomass obtaining costs (including transportation cost) in peri-urban Tokyo is 15% lower than those in mountainous areas (Terada et al. 2010).

After the tsunami-related accident of Fukushima nuclear power plant in 2011, using renewable energy became more widespread and creating distributed local energy supply systems has become an essential need for shaping a resilient society in Japan. Satoyama woodland should not be thought of as simply urban greenery, but as a unique productive landscape that can link ecological restoration and a community energy system.

4.7 The Value of Grey

4.7.1 *Natural Disasters and Layer Model Advantages*

When planning Japanese cities the threat of natural disasters must never be ignored. Only within a recent couple of decades has Japan experienced four major earthquakes and a tsunami; Kobe in 1995, Niigata in 2001, Northeast Japan in 2011, and Kumamoto in 2016. Floods and typhoons also frequently ravage Japanese cities, not only earthquakes and accompanying tsunamis. The *Comprehensive Risk Index* developed by Munich Reinsurance Company (see References) includes all possible risks that cities in the world face, and rates each city. Most cities in Western Europe and North America are a very low number (e.g. Paris 25, London 30, and NYC 42). Compared to European and North American cities, however, Tokyo is an astronomically high 710. This index clearly indicates a fundamental difference in the scale of disaster risk between European and North American cities and those in Japan.

Cities need food. If a distinct boundary exists between urban land use and rural land use and thus the city becomes an entity without agricultural land uses, no food can be generated within its boundary and thus the city will become a place completely dependent on external food supplies. As long as transportation systems are operating normally, cities will avoid any major problems of completely depending on food supplied by rural areas and international markets. However, once a major natural disaster occurs, transportation systems will most likely be seriously damaged, and the external food and energy supply will also most likely be suddenly disrupted. If the city has been completely depending on external food and energy supply, then the loss of transportation systems may inevitably result in the loss of food and energy, and the city will suddenly be caught in a serious situation.

To be prepared for such unpredictable and fatal occurrences, the layer model provides a resilient solution to how land should be planned. Under ordinary conditions, preference can simply be given to the urban layer, and the influence of the rural layer can be minimized. However, when the transportation systems suddenly cease to function because of natural disasters and food supplies have been disrupted, cities shall be able to take advantage of the rural layer and generate its own food inside, or nearby, the city limits. Such a redundant system in food supply based on the layer model, which includes intra- and peri-urban food supplies, may seem inefficient but has the advantage of adaptability to unpredictable changes, and thus highly contributes to a city's resilience. To maintain such a redundant food system, the rural layer should always be embedded in the area as "seeds" to enable immediate response to sudden demands on local food supplies. Japanese cities have realized an increasing probability of such a situation suddenly happening, and the layer model unintentionally maintained high potential to make cities resilient. The advantage of the layer model can be found in its adaptability to a given condition, especially to unpredictable changes such as those caused by natural disasters.

4.7.2 Value of Grey

According to the layer model, the boundary between urban and rural land uses is not as clear as that of a conventional dichotomous model. Also, the boundary should be regarded as in constant flux. The zone between constantly fluxing boundaries may be called a Grey Zone, where an extensive micro-scale mixture of urban and rural land uses is found. In the Grey Zone, the physical entity may change according to changing emphasis on the layers, but the system to control different layers should be there. The key of the layer model is embedded in its intangible system, not in the tangible entity.

Such a system with changing tangible entity controlled by an eternal intangible system can commonly be found in Japan's cultural heritages. Ise Grand Shrine is an excellent example. The Ise Grand Shrine is one of the oldest shrines in Japan, which is well known for more than 1300 years for maintaining a system of rebuilding the shrine buildings every 20 years. Authenticity of the shrine has been embedded in its unique system which has survived over 1000 years, not in its physical entity.

"Grey" stands not for an uncontrolled, uncivilized, or undesirable condition. "Grey" is a keyword that represents an adaptable system where tangible entities may change but the authenticity is embedded in the intangible system itself, and such a system with "grey" character will undoubtedly provide resilience to cities. Usefulness of a planning concept with "grey" can be shared by many cities around the world that also frequently suffer from natural disasters.

Figure 4.41 illustrates the seismic risk hazards and the location of major cities in the world. Red, orange, and yellow show areas with a high risk of seismic hazards; black circles represent the size and location of cities. Many major cities in Asia are obviously situated on terrain where the risk of earthquake is very high. It is expected that not only Japanese cities but cities which share such a high-risk situation will also discover and appreciate the "value of grey" in an effort to make themselves resilient. In recent years, however, such a concern about natural disasters is starting to be shared by cities in the West as well. Because of global climate change, cities along a coastline – no matter where they are located – are now facing the threat of serious storms and coastal flooding, and thus starting to seek an alternative planning concept that may effectively provide them with the needed resilience. The concept of "grey" should also be appreciated not only in Japan but also in the whole world.

Dichotomous landscape with a clear separation of urban and rural land uses is indeed simple, clear, and often beautiful. Such a concept is also efficient provided no sudden or major changes occur. A "grey" landscape with a micro-scale mixture of urban and rural land uses may look chaotic and disordered. However, grey landscape maintains high adaptability to unpredictable and sudden changes, and thus contributes to making cities resilient. "Value of grey" should be appreciated for its resilience and the potential that it holds for the sustainable future of our world.

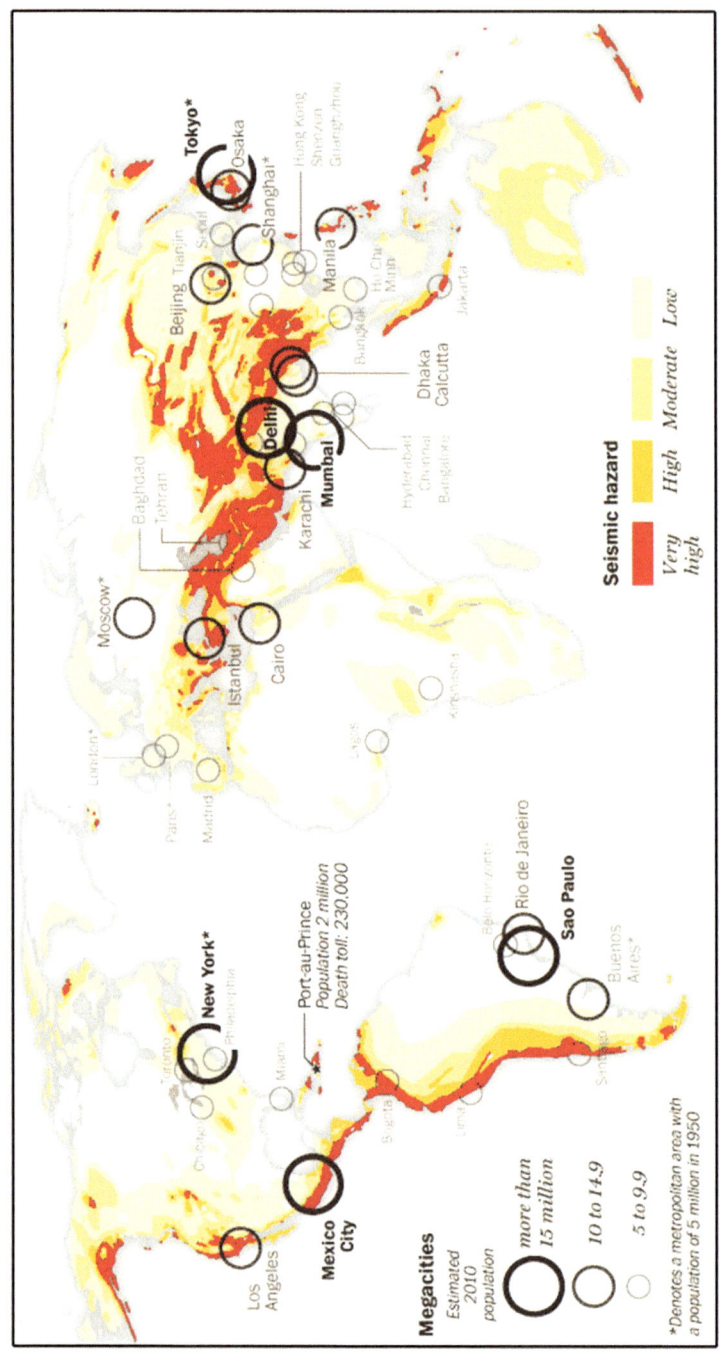

Fig. 4.41 Megacities and seismic hazards in the world. (Source: Global Seismic Hazard Assessment Program, United Nations Population Division)

References

Ashihara Y (1989) The hidden order-Tokyo through the twentieth century. Kodansha USA, New York

Fujii M, Yokohari M and Watanabe T (2002) Edojidai makki no Edo ni okeru nouchi no bunpu jittai no kaimei 江戸時代末期の江戸における農地の分布実態の解明 [Identification of the distribution pattern of farmlands in Edo]. City Planning Review Special Issue, 37, 931–936

Gittleman M, Jordan K, Brelsford E (2012) Using citizen science to quantify community garden crop yields. Cities Environ (CATE) 5(1):4

Matsumoto R, Yokohari M, Terada T, Yamamoto H (2011) Toshi kinkou satoyama ni okeru shimin no kanri ni motozuku kishitsu baio-masu hasseiryou no suitei都市近郊里山における市民の管理にもとづく木質バイオマス発生量の推定 [The amount of harvestable woody biomass from suburban satoyamas by local citizen groups]. J Jpn Inst Landsc Archit 74(5):707–710. (in Japanese with English abstract)

Ministry of Agriculture, Forestry and Fisheries (MAFF) (2016) Heisei 28 nendo shokuryo jukyuuu-hyou (gaisan) 平成28年度食料需給表(概算)[Food balance sheet of FY 2016 (estimates)]. http://www.maff.go.jp/j/zyukyu/fbs/. Accessed 5 May 2017 (in Japanese)

Ministry of Land, Infrastructure, Transport and Tourism (MLIT) (2003) Introduction of urban land use planning system in Japan. https://www.mlit.go.jp/common/000234477.pdf. Accessed 26 Apr 2017

Ministry of the Environment (MOE) (2010) The national biodiversity strategy of Japan. http://www.env.go.jp/en/nature/biodiv/nsj/ (in English). Accessed 6 May 2017

Murayama A (2016) Land use planning for depopulating and aging society in Japan. In: Yamagata Y, Maruyama H (eds) Urban resilience–a transformative approach. Springer, Tokyo, pp 79–92

Murayama A (2017) Urban landscape: urban planning policies and institutional framework. In: Shimizu H, Takatori C, Kawaguchi N (eds) Labor forces and landscape management: Japanese case studies. Springer, Tokyo, pp 61–71

Nakai N (1988) Urbanization promotion and control in metropolitan Japan. Plan Perspect 3(2):197–216

Okata J, Murayama A (2011) Tokyo's urban growth, urban form and sustainability. In: Sorensen A, Okata J (eds) Megacities: urban form, governance, and sustainability. Springer, Tokyo, pp 15–41

Pothukuchi K (2004) Community food assessment: a first step in planning for community food security. J Plan Educ Res 23(4):356–377

Sioen GB, Sekiyama M, Terada T, Yokohari M (2017) Post-disaster food and nutrition from urban agriculture: a self-sufficiency analysis of Nerima Ward, Tokyo. Int J Environ Res Public Health 14(7):748

Statistics Japan (2000) Jinkou suikei 人口推計[Estimated population of Japan]. http://www.e-stat.go.jp/SG1/estat/List.do?bid=000000090004&cycode=0. Accessed 1 May 2017 (in Japanese)

Tahara S, Yokohari M, Kurita H, Terada T (2011) Toshi jumin no nouen ni okeru seisan katsudou ga motarasu nousakubutsu no seisanryou no suitei to sono hyouka 都市住民の農園における生産活動がもたらす農作物の生産量の推定とその評価 [A quantitative assessment of agricultural production from allotment gardens]. J Jpn Inst Landsc Archit 74(5):685–688. (in Japanese with English abstract)

Takeuchi K, Brown RD, Washitani I, Tsunekawa A, Yokohari M (eds) (2012) Satoyama: the traditional rural landscape of Japan. Springer, Tokyo

Terada T (2017a) Urban periphery landscape: dichotomization of urban and rural dimensions. In: Shimizu H, Takatori C, Kawaguchi N (eds) Labor forces and landscape management – Japanese case studies. Springer, Tokyo, pp 73–82

Terada T (2017b) Urban periphery planning: concept to link urban and rural communities in the 21st century. In: Shimizu H, Takatori C, Kawaguchi N (eds) Labor forces and landscape management – Japanese case studies. Springer, Tokyo, pp 381–390

Terada T, Yokohari M, Tanaka N (2010) Shuukaku・yusou kosuto kara mita toshi kinkoubu heich-irin no mokushitsu baiomasu riyou no kanousei 収穫・輸送コストからみた都市近郊部平地林の木質バイオマス利用の可能性 [Potential of utilizing woody biomass from satoyama woods on plateaus assessed by the estimation of harvesting and collecting costs]. Landsc Res Jpn 73(5):663–666. (in Japanese with English abstract)

Tsubota K (2006) Urban agriculture in Asia: lessons from Japanese experience. Food and Fertilizer Technology Center (FFTC), Taipei. http://www.fftc.agnet.org/htmlarea_file/activities/20110719103448/paper-997674935.pdf. Accessed 5 May 2017

Viljoen A, Howe J (eds) (2012) Continuous productive urban landscapes. Routledge, New York

Yagasaki N, Nakamura Y (2010) The role of local groups in the protection of urban farming and farmland in Tokyo. In: Pradyumna K (ed) Local environmental movements. University Press of Kentucky, Lexington, pp 131–114

Chapter 5
Framing in Placemaking When Envisioning a Sustainable Rural Community in the Time of Aging and Shrinking Societies in Japan

Shogo Kudo

Abstract This chapter examines the concept of rural sustainability in the time of an aging and shrinking society. The chapter first introduces the demographic change that Japan is experiencing, a shift from young and growing population to an aged and declining population. Affected by this change, rural regions are facing numerous challenges affecting living conditions of individuals and downscaling socioeconomic activities at regional and communal scales. The multifunctionality framework is applied to understand the past pattern of rural transition. This allows to illustrate subsequent possible phases in the transition driven by an aging and shrinking population. The chapter then provides a review of the placemaking concept, followed by one case study of a placemaking workshop called Monogatari workshop in Gojome, Akita prefecture, Japan. This case study describes how a group of local youth envisioned the future state of their community. The chapter proposes a conceptual illustration of new perspective that the workshop participants gained. The illustration introduces four types of stories, which are story of the past, story of the present, story of the future, and story of oneself. The workshop provided the process to learn personal and collective memories of particular places from older residents of the town. By reflecting on their stories, the participants discussed how they would like to change the same places in the future. The workshop corresponds to the social capital component in the multifunctionality framework which emphasizes intergenerational ties. The chapter suggest the future research should aim to link intergenerational ties to other two capital components of the multifunctionality framework. By doing so, a vision of stable transition to relocalized system will be established even though rural regions continue to experience aging and shrinking of population.

S. Kudo (✉)
Graduate Program in Sustainability Science-Global Leadership Initiative,
Graduate School of Frontier Sciences, The University of Tokyo, Kashiwa, Chiba, Japan
e-mail: kudo@edu.k.u-tokyo.ac.jp

© The Author(s) 2020   97
T. Mino, S. Kudo (eds.), *Framing in Sustainability Science*,
Science for Sustainable Societies, https://doi.org/10.1007/978-981-13-9061-6_5

Keywords Rural sustainability · Placemaking · Aging and shrinking society · Intergenerational ties · Gojome town

5.1 Introduction

5.1.1 Population Aging in Japan and the Challenges in Rural Regions

Japan has been experiencing a shift from a young population to an aged population. This shift is caused by two major demographic changes, namely aging and shrinking population. In 2016, the proportion of people age 65 and older represents 26.7% of the population, the highest figure in the world. The proportion of age 65 and older population is predicted to grow to 39.9% by 2060 (Cabinet Office of Japan 2016). Other evidence of population aging is found in the increase in Japan's median age: 22.2 years in 1950 (Statistics Bureau 2017), and 46.7 years as of 2015. The acceleration of Japan's population aging makes the country's population decline much more prominent. In 2008, the total population of Japan topped at 127.8 million people, the time when Japanese society entered its shrinking phase (Senno 2013).

Recent predictions suggest that the population of Japan will likely decline to 88.1 million people by 2065, a 31.1% decline from the peak population in 2008 (National Institute of Population and Social Security Research 2017). Because increases in life expectancy and decreases in fertility rates are becoming common demographic transitions in not only developed countries but also in developing countries (Harper 2014), many countries are looking at how Japanese society responds to the emerging challenges of an aging and shrinking society.

Along with the aging trend in many countries, considerable differences are found between urban and rural areas. Among developed countries, population aging is more evident in rural areas than urban areas. Only three OECD countries have a higher elderly dependency ratio[1] in urban areas than rural areas: Hungary 28.5%, Poland 23.0%, and Slovakia 20.0% (OECD 2016). In Japan, the core Tokyo metropolitan area has a 4.6% lower proportion of age 65 and older population than the average of other prefectures.[2] This is not caused by a higher fertility rate in Tokyo; in fact, the fertility rate is the lowest in Tokyo, at 1.15 births per woman in 2014

---

[1] Elderly dependency ratio is defined as the proportion of the population aged over 65 to the working-population (commonly defined as the population aged 15–64 in developed countries). Source: OECD 2016.

[2] The core part of the Tokyo metropolitan area consists of Tokyo, Kanagawa, Saitama, and Chiba prefectures. A larger grouping of the Tokyo metropolitan area includes Ibaraki, Gunma, and Yamanashi prefectures (Source: Tokyo Metropolitan Government: http://www.metro.tokyo.jp/ENGLISH/ABOUT/HISTORY/history02.htm). The average proportion of age 65-plus population of the core Tokyo metropolitan area is 24.3%; the average proportion of other prefectures excluding Tokyo metropolitan area and Okinawa is 28.9%. (Source: percentages calculated from Population Census of Japan 2015)

(Cabinet Office of Japan 2016). However, because of the continual migration of youth population from other prefectures, Tokyo retains its lower rate of older people in its population. In the past, the size of rural-to-urban migration was largest during the country's rapid economic growth in late 1950s to early 1970s. Some 7.15 million people migrated to Japan's three major metropolitan areas of Tokyo, Osaka, and Nagoya in the 30 years from 1954 to 1974. Even today Japan continues to observe a similar pattern of migration to the Tokyo metropolitan area, and the high population concentration is raised as the main cause of economic declines in rural regions (Nihon Sousei Kaigi 2014).

5.1.2 Discussing Sustainability in an Aging and Shrinking Phase of Society

Owing to the continuous aging and shrinking of population, Japan's rural regions are facing numerous challenges including those related to living conditions of individual residents (e.g. less accessibility to medical care, infrequent public transportation, loneliness of residents) and the other challenges at the regional societies and communities (e.g. regional economic decline, lack of human resources to assume resource management, community vitality decline). The fact that these challenges are being observed is a sign of social transition from the past phase in which the current economic, social, and political structures were designed to a new phase which requires new social designs. Subsequently, the transition of rural society to an aging and shrinking population phase requires us to re-examine the meaning of sustainability which differs from the topics and scope of sustainability discussed in earlier sustainability science literature.

Since its emergence in the early 2000s, the main focus of sustainability science has been challenges caused by the expansion of human activities such as climate change, land degradation, and energy and resource scarcity. Seminal works of sustainability science state that sustainability science aims to advance our understanding of complex interactions between the ecological system and the human system (Clark 2007; Kates et al. 2001; Komiyama and Takeuchi 2006). Other scholars explain the role of sustainability science in contributing scientific knowledge to sustainable development (Dasgupta 2007; Martens 2006). Aside from these original scopes of the field, sustainability science has the potential of accommodating broader discussions on the normative dimension of the sustainability concept.

Aging and shrinking population poses a question on the way the sustainability concept is understood because the types of challenges these demographic changes deliver are not based on the expansion of human activities. Instead, aging and shrinking of population denote a series of declines in a wide range of social and economic activities. Given the fact that aging and shrinking population will likely remain as a demographic trend of Japanese society in the decades to come, what to sustain becomes unclear.

Because of a greater degree of aging and shirking population, rural regions in Japan are already facing the possible risk of community closures. Earlier studies addressed rural aging and depopulation on different topics such as the decline of agricultural activities (Ishimaru 2009; Sasaki et al. 2007), general living conditions in remote communities (Niinuma 2009; Noguchi et al. 2010; Takegawa 2010), and revitalization of rural economy by increasing interaction with urban residents (Fujita 2005; Tsutsui et al. 2008). However, what the collection of these challenges implies as a larger social transition has not been well discussed. Moreover, accumulating discussions from various case studies on what rural sustainability means and how local residents try to achieve it is essential to better frame a sustainable society for the future.

5.1.3 Aim of This Chapter

This chapter examines the concept of rural sustainability in the time of an aging and shrinking society. This will be achieved by first reviewing the past transition patterns of rural regions based on a multifunctionality framework. This chapter will elaborate how most rural regions have evolved from a farming-based system to a market-based system. The application of the multifunctionality framework allows illustrating subsequent possible phases of rural transition driven by an aging and shrinking population. The chapter then provides a review of the placemaking concept, followed by one case study of a placemaking workshop in the town of Gojome, Akita prefecture, Japan. This case study describes how a group of nine high school students envisioned the future state of their community. The chapter concludes with some discussion on the framings in rural sustainability, and also topics for further studies.

5.2 Rural System Transition: Multifunctionality Framework

One core challenge for the rural regions where aging and shrinking populations are omnipresent is establishing a local system that is capable of coping with various forms of changes in the communities. An analysis of the impact of social changes in individual communities is important because residents experience actual changes and organize concrete reactions in their own communities (Holmes 2006; Wilson 2010). Because aging and shrinking populations are changing the quality of communities, a more comprehensive perspective is required to analyze the entire rural system rather than merely addressing topics individually.

Applying a conceptual framework helps anticipate possible future transitions of rural regions. For this purpose, the literature on rural transformation (Amcoff and Westholm 2007; Ilbery 1998) and urban-rural interactions (Caffyn and Dahlström 2005; Dabson 2007; Phillipson and Scharf 2005; Silverstein et al. 2006; Tacoli 1998) is beneficial. Drawing upon these earlier studies based on systems perspective, this study employs a multifunctionality framework to explore future transitions of rural regions induced by aging and shrinking populations. The following section introduces the multifunctionality discourse, and describes how it functions as framing in discussing rural sustainability.

5.2.1 Conceptual Development of Multifunctionality

The main idea of multifunctionality received a wide range of agreement both from policy makers and academics. Its conceptual development, however, has been diverted and its definition is becoming vague (Andersen et al. 2013; Renting et al. 2009). In many policy-related cases, the definition of multifunctionality has been set individually depending on the purpose of each claim. Regarding this point, Marsden and Sonnino (2008) and Van Huylenbroeck et al. (2007) have provided a comprehensive review on different interpretations of multifunctionality discourse and classified them into three groups: (1) productivist, (2) post-productivist, and (3) sustainable development (Huylenbroeck et al. 2007; Marsden and Sonnino 2008).

The first interpretation of multifunctionality is based on the productivist paradigm which has emerged from a neo-liberalist view of the globalized agricultural market. It realizes the vertical logic and specialization adapted to the globalized market. Multifunctionality in this vision implies the production of multiple outputs from the original inputs provided by primary agricultural production. Such outputs may come in a complementary form to their primary product outputs (Havlik 2005). In this respect, the multifunctional character is limited to the idea of pluriactivity that is formed by agricultural and non-agricultural incomes of farming households (Holmes 2006; Marsden and Sonnino 2008). In the productivist paradigm, individual farming households and the entire agro-food industry are clearly differentiated. Multifunctionality within this paradigm, as a result, is seen as the survival strategy of individual farming households in the global market. As a response to such harsh market circumstances, individual farmers are performing multifunctional agriculture as their coping strategy to keep pace with competition.

The second interpretation of multifunctionality is based on a post-productivist paradigm which focuses on the space of rural areas rather than on production activities. This second interpretation conceives entire rural areas as consumptive targets for amenity demands through eco-tourism, experiencing farm activities, use of rural space for educational purposes, and other means (Barbieri and Valdivia 2010; Huylenbroeck et al. 2007; Marsden and Sonnino 2008; Potter and Tilzey 2005), not only for industrial-based capital. One important actor group in this framing is the urban population who finds scenic and leisure values in rural space. Echoing such normative values of rural areas, the post-productivist view emphasizes environmental protection. Based on this framing, agriculture is perceived as a means of maintaining the local environment of the countryside, not only as a means of food and fiber production. Between the first and second framing in multifunctionality, the conceptual focus shifts from a farm-based approach to a space-based approach through emphasizing the nature and landscape values.

The third interpretation of multifunctionality takes further expansion from the post-productivist paradigm to a sustainable development paradigm. This third view of multifunctionality takes a holistic framing of the concept to realize the connection between socio-environmental benefits from farming operations and the demands of local societies (Marsden and Sonnino 2008). In this view, multifunctionality is viewed as a critical assessment tool for rural development. In contrast to the previous two paradigms in which the social meanings of rural areas are determined by external interests (e.g. food security concerns, competitive agricultural sector, and environmental protection), this third framing interprets multifunctionality as an inclusive development paradigm that takes the internal socioeconomic state into account (Marsden and Sonnino 2008; Morgan et al. 2010). It should be noted that agriculture is seen as one characteristic of rural areas in this third framing. Because the notion of multifunctionality was initially developed in agricultural policies, the multifunctionality discourse has placed its centrality on agriculture, and other dimensions of rural areas were treated as complementary factors. However, such a narrow framing of multifunctionality hinders depicting the rural system transition. In fact, agriculture is no longer the backbone of rural economies today because its proportion in local economic activities has been declining (Huylenbroeck et al. 2007; Milestad and Björklund 2008). Such economic decline triggered an out-migration of young people in search of better employment opportunities in urban areas (Klijin et al. 2005; Milbourne and Doheny 2012).

The author follows the multifunctionality discussions developed by Holms (2006), Marsden and Sonnino (2008), Renting et al. (2009) and Wilson (2008, 2010). In this chapter, multifunctionality is considered as a holistic conceptual framework that illustrates the quality changes in rural areas over time as rural areas undergo a series of social changes. The applied framework in this study is based on the recent works of Wilson (2008, 2010) which depict the quality changes in rural systems in terms of multifunctionality.

In previous empirical studies on multifunctional agriculture, individual farms were used as their analytical unit, and those types of operations that contributed to the multifunctional quality of farmers were examined. These studies commonly set their objectives to quantify the multifunctionality of individual farms by applying parameters (Andersen et al. 2013; Morgan et al. 2010). However, one point of contention is how to select an adequate set of parameters to evaluate the multifunctionality of a target unit quantitatively as the concept of multifunctionality is by no means "clearly and uniformly conceptualized or understood" (Marsden and Sonnino 2008). This view is prominent when a study includes the time-scale because which set of parameters would be appropriate differs according to the socioeconomic state of rural regions over time. Additionally, external factors that possibly affect rural regions are ubiquitous. For example, the state of rural areas before and after the arrival of the global market system should differ from each other considerably.

5.2.2 Development of Multifunctionality Framework

The previous section explained the evolvement of the multifunctionality concept from a framework to evaluate agricultural activities to an all-inclusive concept for rural development. Reflecting the shifts in paradigms over time-from productivist to post-productivist, and to sustainable development-the quality of rural systems illustrates particular transitions, and such transitions represent different qualities of multifunctionality. Based on this idea, Wilson proposes using economic, environmental, and social capitals as a set of descriptive dimensions for depicting diverse qualities of a rural system (Wilson 2008, 2010, 2012). He argues that when the balance of the three capitals is well maintained (well-balanced state in Fig. 5.1), a rural system becomes more stable and achieves self-sustaining capacity. In fact, many rural communities emerged as subsistence farming communities, in which three types of capital were well developed and balanced, and achieved self-managing capacity in producing food, managing local resources, and facilitating minimum exchanges with outside communities.

Illustrating the quality of target systems by sets of indicators has attracted a great deal of research interest across disciplines (Bell and Morse 2008; Morse 2010; OECD 2004; Rametsteiner et al. 2011; Stevens 2005). Wilson provides a solid discussion in studying multifunctionality of rural system with the application of these three types of capital by referring to Bourdieu (1984) to substantiate applying the capital notion (Wilson 2008, 2010). Bourdieu situated capital in three fundamental frames: *economic capital* (material property), *social capital* (networks of social

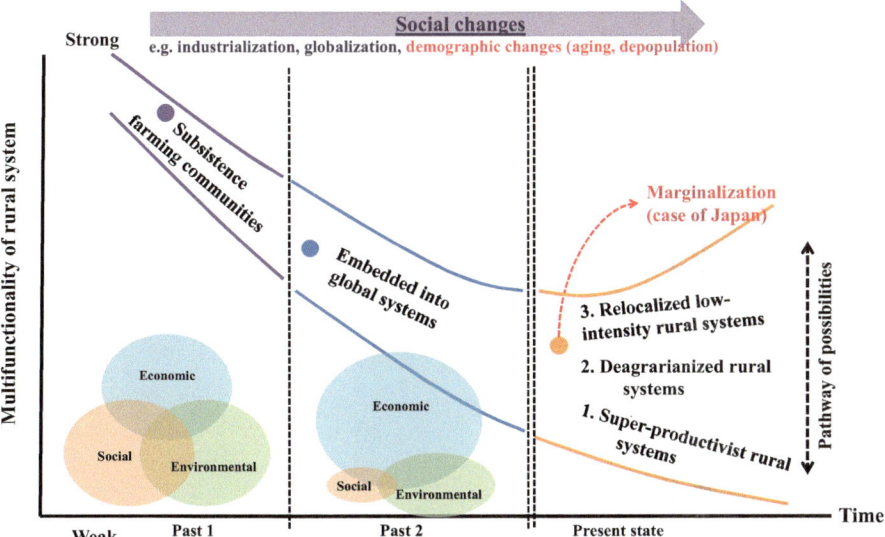

Fig. 5.1 Inter-temporal evolution of the rural system with the quality of multifunctionality. (Source: modified based on Wilson 2008)

connections and mutual obligations), and *cultural capital* (prestige) (Bourdieu 1984). In Bourdieu's theory, the notion of capital is used more as a metaphor or description of processes as individuals or groups gain or lose different types of capital through interactions. Following the approach of Wilson and Bourdieu, this study considers the notion of multifunctionality as a metaphor for analyzing the qualitative changes in rural systems.

Once any concentration on a particular capital is created because of the influence of broader social changes (e.g. economic growth, industrialization, urbanization) or internal changes (e.g. demographic change, political system change, resource scarcity), the configuration of the three capitals is affected. When economic capital is emphasized, and as is often the case, through an industrial change from agricultural production to manufacturing, the quality of a rural system shifts towards the super-productivism which is a state of extremely pronounced economic capital (Wilson 2010). In such a situation, the entire rural system loses the social capital and environmental capital, and the share of economic capital expands largely because the national economy and globalized markets as well as the central government's rural policies directly affect the system (Economic-oriented state in Fig. 5.2).

In another situation, when environmental concerns are realized by certain rural policies oriented toward the post-productivist paradigm (e.g. Agenda 21, CAP scheme), the quality of rural systems moves towards a non-productivism (environmental protection) direction and the environmental capital acquires significant attention (see Environmental-oriented state in Fig. 5.2). If the sustainable development paradigm is reflected properly in the rural development scheme, then a sufficient approach to social capital of local communities should be realized; ideally this approach should usher a rural area into the well-balanced state as shown in Fig. 5.2.

The illustrations of the different qualities of multifunctionality based on economic, environmental, and social capital can be applied to an inter-temporal system

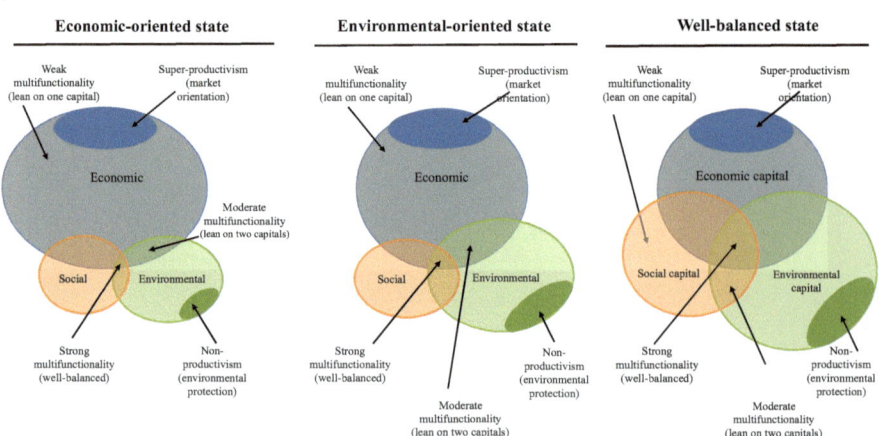

Fig. 5.2 Different quality of multifunctionality based on economic, social and environmental capitals. (Source: modified from Wilson 2010)

transition model as shown in Fig. 5.1. Venn images are added to represent possible configurations of three capitals that describe each state of rural system transition. The figure depicts two past states of rural system and possible future pathways based on the present. The focal unit of this framework is a rural region and the quality of multifunctionality in each state is described by the totality of all kinds of local resources, attributes of individual residents, and activities taking place. The separations of three states indicate each time period. However, these distinctions are unclear because a system transition occurs over a long time, anywhere from 50 to 100 years perhaps, and is not spontaneous unless a fast and sudden rupture changes the state of system quality dramatically.

The initial state of rural system is described as "Past 1. Subsistence farming" in Fig. 5.2. In this phase, rural communities are agrarian and self-sufficient in food and generally energy production. The quality of multifunctionality, therefore, appears strong. The three types of capital are considered well balanced at this stage. Such a solid balance is a crucial condition for the survival of a subsistence community.

Once rural communities increase their engagement with the outside world, often when the global market is introduced, a rural system transits to "Past 2. Embedded into global systems" phase (Fig. 5.2). In this phase, the economic capital strengthens to build an economical-oriented system. This transition from subsistence farming to an embedded system tends to drive agriculture toward intensification and monoculture-based production, and causes declines in environmental and social capital because of degradation in the local environment and out-migration of young people (Parnwell 2007; Rigg et al. 2008; Wilson 2010).

The present state, the third state in Fig. 5.2, presents three possible pathways: (1) super-productivist rural systems, (2) deagrarianized rural systems, and (3) relocalized low-intensity rural systems (Wilson 2010). Super-productivist rural systems imply the economic-oriented state in Fig. 5.1, in which the concentration on economic capital is pursued. Towards this direction, approaches based on the productivist paradigm such as intensification and specialization in agricultural production are preferred. The pathway of a deagrarianized rural system initiates a transition to non-agricultural sectors in rural areas. The main goal of deagrarianization is achieving alternative measures to ensure economic capital development in rural systems. The third possible pathway, a relocalized low-intensity rural system direction, aims at a higher multifunctionality quality than the other two pathways by achieving a well-balanced relationship among the three types of capital.

Based on this discourse, the choice of '1' would lead to a lower quality of multifunctionality in a rural system, and '2' and '3' would maintain the same or higher quality. However, these three possible pathways are not separated completely, and in reality, each rural area or community would exhibit the mixed directions through 1–3. At the same time, all rural areas would still be affected largely by the external, broader interests of society. For example, individual communities can direct themselves to a relocalized rural system pathway by conducting local initiatives, whereas the national rural policy can be oriented towards super-productivist to build a competitive agricultural sector in the global market. In fact, the introduction of a market mechanism caused a transition from subsistence farming (Past 1 state in Fig. 5.2) to one embedded in the global system (Past 2 state in Fig. 5.2) by increasing economic capital. Towards the future, other changes in larger systems such as national policy reforms, globalization, and demographic changes would predetermine the possible directions of system transition. The possible transitional space is termed as *pathway of possibilities* as shown in Fig. 5.2 (Wilson 2008).

Apart from the original three pathways, the marginalization pathway is becoming more and more realistic as the fourth possibility for today's rural regions in Japan (Fig. 5.2). The marginalization denotes an excessive degree of qualitative decline in community vitality as well as collective actions of residents in the rural communities (Kudo and Yarime 2013). Community vitality is a relatively new concept and it refers to the ability of a community to "sustain itself into the future as well as provide opportunities for its residents to pursue their own life goals and the ability of residents to experience positive life outcomes" (Crandall and Etuk 2008).

As a rural area enters this marginalization pathway, residents experience declines in various aspects in living conditions such as mobility and access to basic items (Asai et al. 2012; Kuramochi et al. 2014), management of vacant houses and abandoned farmlands (Ishimaru 2009; Shinobe and Miyachi 2012; Yamamoto and Nakazono 2008), and seasonal events and daily chores such as grasscutting in cemetery areas, removing mud and leaves clogging water channels, and cleanups along residential roads (Niinuma 2009; Tamasato 2009). These declines in the communities are considered possible threats to the well-being of residents.

Today, community marginalization is expanding to the small-size municipalities such as villages and towns in Japan, and has ceased to be an issue in rural communities. Knowing aging and shrinking population will most likely drive a rural system to its next transition, it is becoming critical for rural residents to take their own initiatives in responding to the emerging challenges. To implement such initiatives, rural residents must discuss and envision the future they want to have in their own communities. The following section introduces a placemaking workshop conducted with a group of local youths in Gojome, Akita prefecture, Japan, as an example case of such local initiative.

5.3 Envisioning a Sustainable Community in an Aging and Shrinking Society: Case of *Monogartari* Workshop in Gojome Town

5.3.1 Context

Gojome is located in central Akita (Fig. 5.3). The town is largely agrarian but also known for its forestry and timber processing industry. The population is 9269 and the proportion of residents age 65 and older is 42.4% as of October, 2016. Akita has been known as the prefecture with the highest proportion of population age 65 and

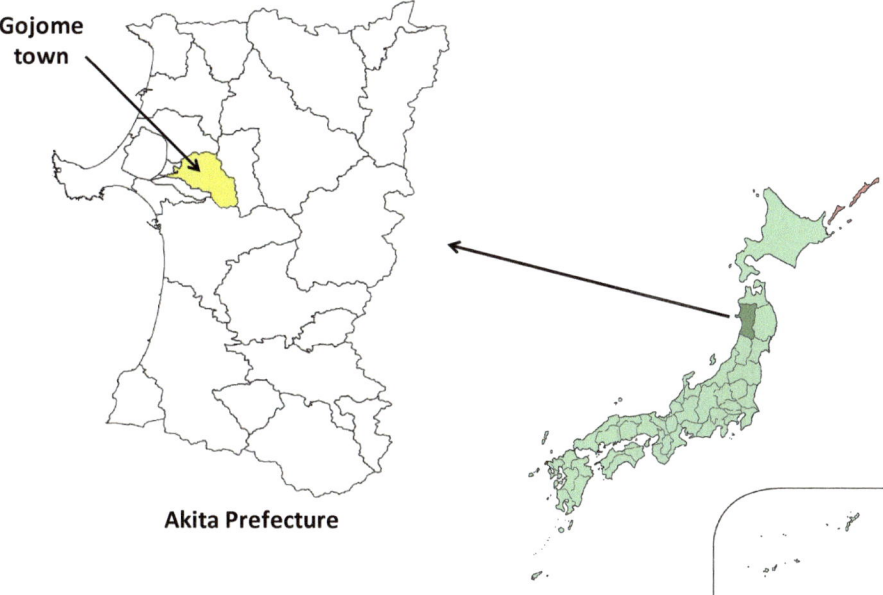

Gojome town

Akita Prefecture

Fig. 5.3 Location of Gojome in Akita prefecture, Japan

over, which was 34.4% in 2016, and the highest depopulation rate in the last three census surveys. Gojome is a representative small-size municipality in rural Japan where aging and shrinking population have influenced local communities in terms of its economic, environmental, and social activities.

Figure 5.4 shows population changes of Gojome for 1965–2015, and population predictions until 2040. In 1965, the total population of Gojome was 18,862 people. The town has experienced continual population decline since 1965, and in 2015 the total population was 9433 people, about 50% of the total population size of 1965 (Fig. 5.4). Subsequently, the share of young people (age 0–15) declined over time: around 20% in 1980, but only 7.9% of the total population today. This continuous decline of local youth further accelerates the aging of Gojome.

One main reason for the town's continuous and constant depopulation is the out-migration of local youth, particularly 15- to 18-year-olds. According to the principal of Gojome Junior High School, the only junior high school in Gojome, only 30% of graduates choose to go Gojome High School, the only high school in Gojome. The remaining graduates go to high schools in neighboring municipalities. The author and, Ryu Yanagisawa, community development officer in Gojome, had a discussion with the teachers of Gojome High School and shared a concern that the type of education local schools have provided might have influenced the local youth's mindsets to naturally imagine their better future would always be outside the town.

We also shared a concern that the town is losing its next generation of residents who play a key role in sustaining local activities and traditions. The author, Mr.

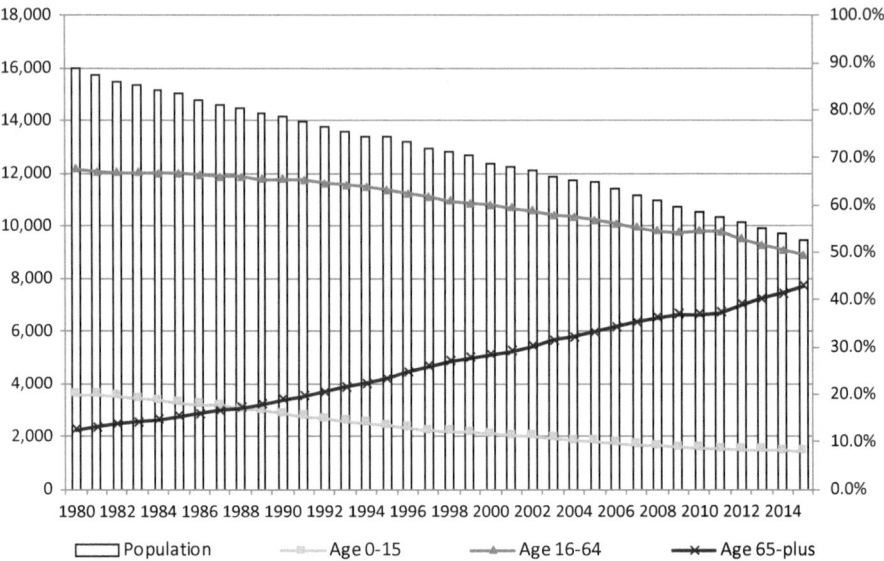

Fig. 5.4 Population change of Gojome since 1965. (Data source: Gojome Town Population Vision, 2016)

Yanagisawa, and three teachers from Gojome High School agreed that local youth are an important focus group for the town to train human resources who take initiatives to design the future of Gojome. Based on this common understanding, a workshop targeting a group of nine local youth in the town was conducted.

5.3.2 Conceptual Design of Workshop: Placemaking Concept

To develop a workshop for the local youth, the concept of placemaking was utilized. Placemaking has emerged in urban studies and is a process that designs a third place in urban environment. *Third place* is a concept coined by American urban sociologist Ray Oldenburg and refers to places inside cities where people feel comfortable to be and to socialize with others, which are separately recognized from their home (first place) and their work environment (second place) (Oldenburg 1999). Some examples of third places are cafés, libraries, and art galleries. Oldenburg's argument is that cities need the function of a third place to ensure the well-being of residents. However, in reality, the process of placemaking extends beyond the physical planning of third places; it includes various forms of discussions and negotiations among different stakeholders both in its planning and the actual operation of such third places.

Mitomo (2015), a leading scholar on placemaking concept in Japan, conducted a review on the placemaking concept. Although the covered literature was limited to books and reports, she concluded that the concept is ill-defined and concrete implementation steps have yet to be well presented (Mitomo 2015). The author extended the literature review on the placemaking concept to academic journals. A simple online search[3] found more than 580 journal articles that have "placemaking" in either titles, abstracts, and keywords; however, only a limited number of papers provide clear definitions of the concept. Instead, many of them are empirical studies targeting specific challenges of their case study areas located in different geographical areas. For example, placemaking is applied in studies on community energy projects to understand the social perception and residents' acceptance to energy facility installations (Fast and Mabee 2015; Middlemiss and Parrish 2010).

The concept is also discussed as a possible means for organizing groups of people for social movements (Larsen 2008; Lepofsky and Fraser 2003). Combined with text analysis on social media, placemaking is suggested as a method to realize virtual landscapes in actual city designs (Alkadri et al. 2015). Overall, the review of earlier literature suggests two types of definitions: the conceptual definition, and the working definition.

Regarding the conceptual definition, Pierce et al. (2011) provides a comprehensive definition of placemaking: "the set of social, political and material processes by which people interactively create and recreate the experienced geographies in which they live" (Pierce et al. 2011). In their definition placemaking goes beyond physical planning and requires people's frequent participation in the process. Placemaking is also explained as local responses of the residents of the specific area where particular social issues are present. To structure socially fair and operationally sustainable responses, those particular social issues and places related to the residents need to be framed properly. This process of placemaking is fundamentally linked with how people frame their living environment.

In earlier literature in communication and political science, frame or framing is understood as how individuals organize their experiences and make sense of social events that they encountered (Benford and Snow 2000; Goffman 1974). Framing, at the same time, contributes to recognizing a controversy that resonates with people's core values and assumptions (Nisbet and Mooney 2007). Hence, the placemaking process includes positive, neutral, and negative interpretations of particular social events and associated places. Additionally, along with its affiliation to individuals, framing is generated as the result of collective organizational narratives that reflect the cultural values of people (Benford 1997).

Considering neighborhood and communities as its focal point, placemaking is fundamentally a collective process by a group of residents. Through framing, people judge what issues are relevant to them and what issues are not, who should be responsible for the issues, and what actions should be taken (Gamson et al. 1992; Price et al. 2005). Through this process, placemaking guides us to re-examine the framings that people hold of living environment through organizing a

---

[3] This search was conducted on 20 November 2016 in Science Direct and Google Scholar. The search was conducted by using "placemaking" as the keyword specified in title, abstract, and keyword list. The search resulted in 581 hits in Science Direct and 588 hits in Google Scholar.

series of dialogues among residents and collectively envision a desirable future state of community.

Regarding the working definition, Placemaking Chicago, a well-respected NGO involved in urban planning, provides a clear definition of placemaking:

> "Placemaking is a people-centered approach to the planning, design and management of public spaces. Put simply, it involves looking at, listening to, and asking questions of the people who live, work and play in a particular space, to discover needs and aspirations. This information is then used to create a common vision for that place. The vision can evolve quickly into an implementation strategy, beginning with small-scale, do-able improvements that can immediately bring benefits to public spaces and the people who use them." (*under-lining added by the author for emphasis.)

Project for Public Spaces based in New York, another well-respected organization working on urban development, adds that placemaking is to "inspire people to collectively reimagine and reinvent public spaces as the heart of every community" (Project for Public Space 2016). Several other NGOs based in English-speaking countries work for the implementation of placemaking concept to solve local challenges. All of these organizations emphasize the participation of residents in the process of placemaking.

By integrating the conceptual definition of scholars and the working definition of practitioners, this study defines placemaking as the collective social, political, and material processes planned and implemented by a group of people who reside in a particular place to envision the future state of their communities. These processes include a series of formal and informal gatherings in which a wide range of topics and methods to realize their desirable form of community are discussed.

5.3.3   Practice: Monogatari Workshop

The workshop was organized for 2 days in August 2016. By applying a Japanese expression of "stories" (*monogatari*), the workshop was entitled *Monogatari Workshop*. The objective was to elicit stories about places in the town of Gojome about which people have personal or collective memories. This workshop corresponds to the beginning of placemaking process, re-examining people's understanding about the environment in which they live in. Nine students from Gojome High School participated, including two local residents: one man in his early 50s and another man in his early 60s served as key informants for the interviews included in the workshop.

Photography was utilized as the main method of this workshop. Firstly, old pictures taken 50–100 years ago in Gojome were used during the interviews to enable the informants to recount stories of the past (Fig. 5.5). By asking key informants questions about the events, items, and activities shown in those pictures, the participants learned the lifestyle of people in the past (Fig. 5.6). The two adult residents

a   b

Fig. 5.5 (**a**) Street market (ca. 1930). (**b**) Street market (2016)

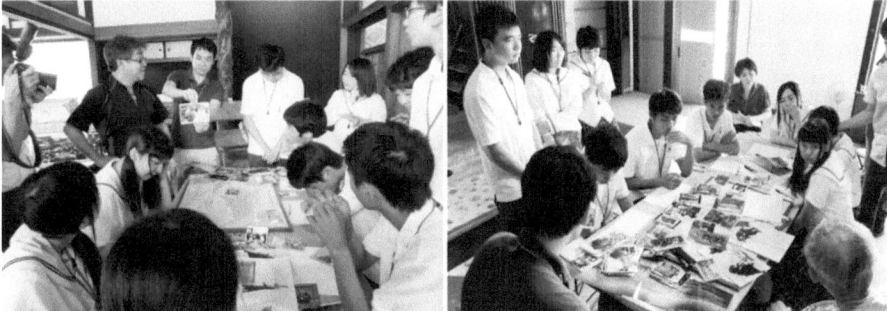

Fig. 5.6 Interviewing two local residents

who participated as key informants were owners of a bookstore and a photo studio. Some of the workshop participants found the pictures of the local farmer's market interesting because many changes were identified such as people's clothing, items they were selling, and the crowdedness of the market.

Secondly, after learning the stories of the past from the interviews, the participants each chose one picture and visited the same location to observe buildings, landscape, and people's activities of today. By identifying landmarks at each place such as old walls, power line poles, and mountain scenery, the participants took pictures from an angle that best approximated the same angle in the old pictures. They also walked around the place in the old pictures and observed the activities taking place today. When encountering some people in the area, they interviewed them to collect information about those places. Through this process, the participants gained a better understanding of the places they had chosen in two different points of time. This activity helped the participants perceive the Gojome of 2016

Fig. 5.7 Reporting findings and final exhibition

based on how places were used by people in the past, not only on how places look like today. Such in-depth understanding about places in their town allowed them to re-examine the value of places today.

Thirdly, after taking the picture of today at the same locations as in the old pictures the participants had chosen, the participants shared the stories of the past and the present. They described the old pictures they chose, shared what they found interesting in the old pictures, and how they felt by visiting those places. By utilizing old pictures and the pictures participants took by themselves on the day of workshop, the participants discussed what kind of community they want to actualize in the future. This last process corresponds to the process to collectively envision the future state of community in placemaking (Fig. 5.7). Pictures taken by the workshop participants and the old pictures used during the interview were exhibited with some explanatory notes at a local community center. This exhibition was for Gojome residents who wanted to learn the changes in places in the town and who wanted to join the placemaking process.

Through this workshop, the participants obtained a new perspective to envision their future through learning three stories about their local places and sharing his or her fourth story with other participants. The first one is the *story of the past* from the interviews. This process raised awareness of the participants regarding how places looked in the past and what kinds of activities were conducted then.

The second narrative is the *story of the present*. The picture-taking activity and interviews at the same location as in old pictures provided multiple understandings about local places. All participants mentioned that they paid little attention to how places have changed over time prior to this workshop. By learning those changes from the past to the present, participants gained the story of today.

The third story provided insights into what may come, the *story of the future*. This was created collectively by workshop participants through sharing their ideas for the desirable future state of the places they have learned through the workshop.

Additionally, we also discussed what individual participants can do to actualize the envisioned future of local communities. Interestingly, most of the participants expressed that they would like to preserve the current landscape, features, and activities of the places they studied for their future instead of creating new activities.

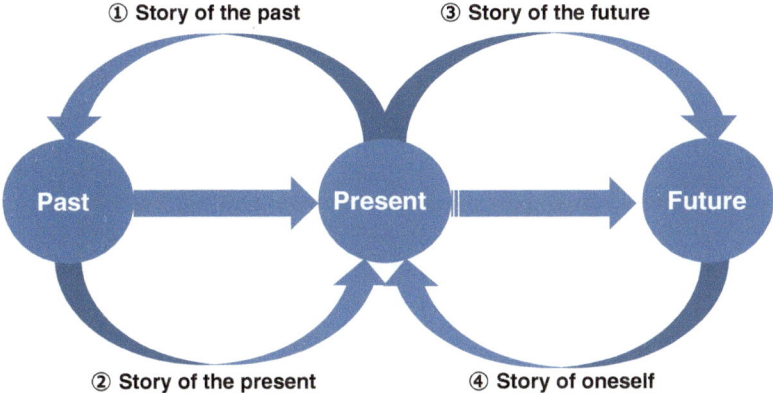

Fig. 5.8 Participants new perspective gained through Monogatari workshop

Finally, as the fourth story, each participant reported how they would like to react to the third story they discussed. Summarizing each participant's report to develop a concrete action plan is not the intention of this step. Instead, acknowledging others' ideas for actual reactions to what they learn is the goal of this last step. Figure 5.8 summarizes the perspective participants gained through this workshop. This infinity-symbol-like shape perspective illustrates a retrospective view when envisioning the future based on the present and the past. Instead of directly perceiving the present from the past and anticipating the future from the present, workshop participants took four stories as steps to deepen their understanding about the past, the present, and a possible Gojome of the future. The fourth story they reported was the *story of oneself*, which inquires what the participant would do as an individual member of the local community at the present moment to actualize the discussed desirable future through workshop.

5.4 Concluding Discussion

Reflecting on the previous sections, two major discussion points on framing involved in rural sustainability in an aging and shrinking society arose. The first point is a macro-scale framing in rural transition and how a rural region can be directed to a sustainable development pathway. The multifunctionality framework described past transition patterns of rural systems and provided three possible directions of system transition pathways driven by various social changes. The past transitions were largely caused by global market mechanisms and the increased awareness of environmental protection. The sustainable development notion is claimed as a recent driver of rural system transition; however, its reflection of reality is often limited to the policy level such as the Common Agricultural Policy (CAP) scheme in Europe.

   This chapter argues that aging and shrinking population has emerged as prominent driver of system transition and is accompanied by the fourth possible direction of transition, the marginalizing pathway. Rural areas of Japan are the most typical cases where such transition is taking place. The emergence of marginalization pathways is inevitable because the total population of the country is in a long-term trend of an aging and shrinking population. When such a declining pathway is suggested, the meaning of rural sustainability becomes increasingly nebulous and must be redefined.

   In extension of the first point, the second discussion point concerns the process for deciding what to sustain in rural communities in the time of aging and shrinking societies. This is a critical framing issue on what people as a society aim to sustain in rural regions and what rural residents want to sustain in their own living environment. Multifunctionality framing notes that scenic and leisure values in natural landscape, and historical and cultural values, are to sustain in rural regions. This argument has been the main justification for rural policies that provide a wide range of support to rural regions, particularly through agricultural policies, both in Europe and Japan. However, this line of discussion is rather top-down and situates rural residents as the stewards of the regional assets, which national policies consider important.

   More importantly, the voices of rural residents are not well-reflected and the ownership of discussion is missing. Placemaking serves as a bottom-up approach to initiate discussion among rural residents to redefine their understanding to living environment and to identify what residents want to sustain in their own communities. Furthermore, the placemaking process guides the participants to collectively envision a desirable state of their own communities in the future. This participatory process generates a new mindset among the participants to engage with local issues. This chapter introduced a *Monogatari* placemaking workshop in Gojome, Akita prefecture, Japan which trained nine high school students in the town to have a retrospective view to reframe their understanding of local places and history. Such an interactive process is an integral part of the core of placemaking and helps the participants commit themselves to the ownership of present-day local challenges.

   To conclude, discussing rural sustainability in an aging and shrinking phase of society emphasizes the importance of intergenerational ties. In such aging and shrinking communities in rural Japan like Gojome, identifying a method to lead their community to the relocalized low-intensity rural systems direction is critical (see "3. Relocalized low-intensity rural systems" in Fig. 5.2) because this can ensure the highest degree of multifunctionality. This is also desirable because the most stable state of a rural system is achieved. This study suggests placemaking as one method to lead a rural region towards this direction. The workshop contributed to creating intergenerational ties about different places in Gojome through learning four different stories namely story of the past, story of the present, story of the future, and story of oneself (Fig. 5.8).

   The workshop provided the process to learn personal and collective memories of particular places from two local residents in different generations. By reflecting on their stories, the participants discussed how they would like to change the same places in the future. This practice addressed the social capital component in the multifunctionality framework. The next step for the author is to expand intergenerational ties to the other two capitals in the multifunctionality framework. By doing so, a vision of stable transition to relocalized system will be established among the local residents even though rural regions continue to experience aging and shrinking of population.

References

Alkadri MF, Istiani NFF, Yatmo YA (2015) Mapping social media texts as the basis of place-making process. Procedia Soc Behav Sci 184:46–55. https://doi.org/10.1016/j.sbspro.2015.05.052

Amcoff J, Westholm E (2007) Understanding rural change demography as a key to the future. Futures 39(4):363–379. https://doi.org/10.1016/j.futures.2006.08.009

Andersen PS, Vejre H, Dalgaard T, Brandt J (2013) An indicator-based method for quantifying farm multifunctionality. Ecol Indic 25:166–179. https://doi.org/10.1016/j.ecolind.2012.09.025

Asai H, Kawasuso T, Oura F (2012) Tottori-ken ni okeru Kaimono Jyakusha Taisaku ni kansuru Torikumi – Chukan Chiiki no Kaimono Jyakusha Taisaku ni kansuru Kisoteki Kenkyuu Sone 1 [鳥取県における買い物弱者対策に関する取り組み–中間地域の買い物弱者対策に関する基礎的研究その1]. J Arch Build Sci 1:163–164

Barbieri C, Valdivia C (2010) Recreation and agroforestry: examining new dimensions of multifunctionality in family farms. J Rural Stud 26(4):465–473. https://doi.org/10.1016/j.jrurstud.2010.07.001

Bell S, Morse S (2008) Sustainability indicators measuring the immeasurable, 2nd edn. Earthscan, Abingdon

Benford RD (1997) An insider's critique of the social movement framing perspective. Sociol Inq 67(4):409–430. https://doi.org/10.1111/j.1475-682X.1997.tb00445.x

Benford RD, Snow DA (2000) Framing processes and social movements: an overview and assessment. Annu Rev Sociol 26:611–639. Retrieved from http://www.jstor.org/stable/223459

Bourdieu P (1984) Distinction: a social critique of the judgement of taste. Routledge, London

Cabinet Office of Japan (2016) Kourei Shakai Hakusho (White paper report on the aging society 2016 高齢社会白書2016). Tokyo. Retrieved from http://www8.cao.go.jp/kourei/whitepaper/w-2016/zenbun/28pdf_index.html

Caffyn A, Dahlström M (2005) Urban-rural interdependencies: joining up policy in practice. Reg Stud 39(3):283–296. Retrieved from http://www.tandfonline.com/doi/abs/10.1080/0034340050086580

Clark WC (2007) Sustainability science: a room of its own. Proc Natl Acad Sci U S A 104(6):1737–1738. https://doi.org/10.1073/pnas.0611291104

Crandall M, Etuk L (2008) What is community vitality? Retrieved August 17, 2015, from http://oregonexplorer.info/content/what-community-vitality

Dabson B (2007) Rural-urban interdependence: why metropolitan and rural america need each other. The Brookings Institution, Washington, DC

Dasgupta P (2007) The idea of sustainable development. Sustain Sci 2(1):5–11. https://doi.org/10.1007/s11625-007-0024-y

Fast S, Mabee W (2015) Place-making and trust-building: the influence of policy on host community responses to wind farms. Energ Policy 81:27–37. https://doi.org/10.1016/j.enpol.2015.02.008

Fujita Y (2005) Komyuniti Bijinesu ga Kirihiraku Chiiki-zukuri [The potential for regional development through community business コミュニティビジネスが切り開く地域づくり]. Tottori Univ J Fac Reg Sci 2(1):11–27

Gamson WA, Croteau D, Hoynes W, Sasson T (1992) Media images and the social construction of reality. Annu Rev Sociol 18:373–393

Goffman E (1974) Frame analysis: an essay on the organization of experience. Northeastern University Press, Boston

Harper S (2014) Ageing societies. Routledge, London

Havlík P (2005) Joint production under uncertainty and multifunctionality of agriculture: policy considerations and applied analysis. Eur Rev Agric Econ 32(4):489–515. https://doi.org/10.1093/erae/jbi027

Holmes J (2006) Impulses towards a multifunctional transition in rural Australia: gaps in the research agenda. J Rural Stud 22(2):142–160. https://doi.org/10.1016/j.jrurstud.2005.08.006

Huylenbroeck G Van, Vandermeulen V, Mettepenningen E, Verspecht A (2007) Multifunctionality of agriculture: a review of definitions, evidence and instruments. Living Rev Lands Res 1(3). https://doi.org/10.12942/lrlr-2007-3

Ilbery B (1998) The geography of rural change, 2nd edn. Routledge, Abingdon, Oxon

Ishimaru N (2009) Genkai shuraku to iwarete iru shuraku ni okeru 耕作放棄地に関する研究 [A study on the situation of abandoned cultivated land in the village, so-called squeezed village – in a case of Soradani district, Aki-Oota-cho, Hiroshima Prefecture]. Architect Inst Jpn NII Electron Libr Serv 32:4

Kates RW, Clark WC, Corell R, Hall JM, Jaeger CC, Lowe I, McCarthy JJ, Svedin U (2001) Sustainability science. Science 292(5517):641–642. https://doi.org/10.1126/science.1059386

Klijn JA, Vullings LAE, Lammeren RJA van, Meijl H van, Rheenen T van, Veldkamp A, Verburg PH, Westhoek H, Eickhout B, Tabeau AA (2005) The EURURALIS study: technical document. Alterra. Retrieved from http://library.wur.nl/WebQuery/wurpubs/343642?wq_sfx=wever

Komiyama H, Takeuchi K (2006) Sustainability science: building a new discipline. Sustain Sci 1(1):1–6. https://doi.org/10.1007/s11625-006-0007-4

Kudo S, Yarime M (2013) Divergence of the sustaining and marginalizing communities in the process of rural aging: a case study of Yurihonjo-shi, Akita, Japan. Sustain Sci 8(4):491–513. https://doi.org/10.1007/s11625-012-0197-x

Kuramochi H, Tanimoto K, Tsuchiya S (2014) 中山間Chiiki ni okeru Kaimono Shienni kansuru Kosatsu – Ido Hanbai ni Kosatsu – Ido Hanbai ni Chumoku shite – (Shopping support in rural depopulated areas: focusing on delivery services). Sociotechnol Res Netw 11(April):33–43

Larsen SC (2008) Place making, grassroots organizing, and rural protest: a case study of Anahim Lake, British Columbia. J Rural Stud 24(2):172–181. https://doi.org/10.1016/j.jrurstud.2007.12.004

Lepofsky J, Fraser JC (2003) Building community citizens: claiming the right to place-making in the city. Urban Stud 40(1):127–142. https://doi.org/10.1080/00420980220080201

Marsden T, Sonnino R (2008) Rural development and the regional state: denying multifunctional agriculture in the UK. J Rural Stud 24(4):422–431. https://doi.org/10.1016/j.jrurstud.2008.04.001

Martens P (2006) Sustainability: science or fiction? Sustain Sci Pract Pol 2(1):36–41

Middlemiss L, Parrish BD (2010) Building capacity for low-carbon communities: the role of grassroots initiatives. Energ Policy 38(12):7559–7566. https://doi.org/10.1016/j.enpol.2009.07.003

Milbourne P, Doheny S (2012) Older people and poverty in rural Britain: material hardships, cultural denials and social inclusions. J Rural Stud 28(4):389–397. https://doi.org/10.1016/j.jrurstud.2012.06.00

Milestad R, Björklund J (2008) Strengthening the adaptive capacity of rural communities: multifunctional farms and village action groups. In: Proceedings of the 8th European IFSA symposium, July, pp 6–10. Retrieved from http://ifsa.boku.ac.at/cms/fileadmin/Proceeding2008/2008_WS3_02_Milestad.pdf

Mitomo N (2015) Puresumeikingu no teigi gensoku to ba no hyouka koumoku ni kansuru kousatsu. [プレスメイキングの定義原則と場の評価項目に関するる察] 日本デザイン学会デザイン学研究

Morgan SL, Marsden T, Miele M, Morley A (2010) Agricultural multifunctionality and farmers' entrepreneurial skills: a study of Tuscan and Welsh farmers. J Rural Stud 26(2):116–129. https://doi.org/10.1016/j.jrurstud.2009.09.002

Morse S (2010) Sustainability: a biological perspective. Cambridge University Press, Cambridge

National Institute of Population and Social Security Research (2017) 日本の将来推計人口（平成29年推計）. Tokyo. Retrieved from http://www.ipss.go.jp/pp-zenkoku/j/zenkoku2017/pp29_gaiyou.pdf

Nihon sousei kaigi [Japan Policy Council] (2014) Stoppu shoushika chihou genki senryaku (ストップ少子化・地方元気戦略 Stop depopulation strategies for regional vitality). Tokyo. Retrieved from http://www.policycouncil.jp/

Niinuma S (2009) Considerations on the maintenance of village functions and the sustainability of residents' daily lives at "marginal settlements": the case of settlement M of Hinohara-Mura, Tokyo. E J GEO 4(1):21–36. Retrieved from http://wwwsoc.nii.ac.jp/ajg/ejgeo/412136Niinuma.pdf

Nisbet MC, Mooney C (2007) Framing science. Science 316(5821):56. https://doi.org/10.1126/science.1142030

Noguchi K, Sato S, Kobayashi Y, Himeno Y, Siiba N, Terada M (2010) Classification and characteristics of village by living environment evaluation -living environment and the sphere in Saiki City, Part1- (in Japanese). The Japan Association of Economic Geographers, pp 3–4. Retrieved from http://ci.nii.ac.jp/els/110008112674.pdf?id=ART0009637691andtype=pdfandlang=jpandhost=ciniiandorder_no=andppv_type=0andlang_sw=andno=1335239391andcp=

OECD (2004) Measuring sustainable development. Paris. Retrieved from http://www.oecd.org/site/worldforum/33703829.pdf

OECD (2016) Population and migration estimates. OECD Factbook. OECD. https://doi.org/10.1093/oxfordhb/9780199589531.013.0035

Oldenburg R (1999) The great good place: cafes, coffee shops, bookstores, bars, hair salons, and other hangouts at the heart of a community, 3rd edn. Marlowe and Company, New York

Parnwell MJG (2007) Neolocalism and renascent social capital in Northeast Thailand. Environ Plann D 25(6):990–1014. https://doi.org/10.1068/d451t

Phillipson C, Scharf T (2005) Rural and urban perspectives on growing old: developing a new research agenda. Eur J Ageing 2(2):67–75. https://doi.org/10.1007/s10433-005-0024-7

Pierce J, Martin DG, Murphy JT (2011) Relational place-making: the networked politics of place. Trans Inst Br Geogr 36(1):54–70. https://doi.org/10.1111/j.1475-5661.2010.00411.x

Potter C, Tilzey M (2005) Agricultural policy discourses in the European post-Fordist transition: neoliberalism, neomercantilism and multifunctionality. Prog Hum Geogr 29(5):581–600. https://doi.org/10.1191/0309132505ph569oa

Price V, Nir L, Cappella JN (2005) Framing public discussion of gay civil unions. Public Opin Q 69(2):179–212. https://doi.org/10.1093/poq/nfi014

Project for Public Spaces (2016) Placemaking: what if we built our cities around places? New York. Retrieved from http://www.pps.org/wp-content/uploads/2016/10/Oct-2016-placemaking-booklet.pdf

Rametsteiner E, Pülzl H, Alkan-Olsson J, Frederiksen P (2011) Sustainability indicator development – science or political negotiation? Ecol Indic 11(1):61–70. https://doi.org/10.1016/j.ecolind.2009.06.009

Renting H, Rossing WAH, Groot JCJ, Van der Ploeg JD, Laurent C, Perraud D, Stobbelaar DJ, Van Ittersum MK (2009) Exploring multifunctional agriculture. A review of conceptual approaches and prospects for an integrative transitional framework. J Environ Manag Suppl 2(90):112–123. https://doi.org/10.1016/j.jenvman.2008.11.014

Rigg J, Veeravongs S, Veeravongs L, Rohitarachoon P (2008) Reconfiguring rural spaces and remaking rural lives in Central Thailand. J Southeast Asian Stud 39(3):355–381. https://doi.org/10.1017/S0022463408000350

Sasaki H, Koyama K, Matsuura S (2007) 耕作放棄地の分布と潜在生産力の推定(Geographical distribution and potential grass productivity of abandoned cultivated land in Japan9). Jpn Soc Grassland Sci 53(3):189–194

Senno M (2013) 人口減少社会「元年」は, いつか? Jinkou genshou shakai 「gannen」 ha, itsu ka? (When was the first year of population declining society?) Retrieved January 10, 2014, from http://www.stat.go.jp/info/today/009.htm

Shinobe H, Miyachi T (2012) Akiya no の解体除去施策の現状と課題-西日本の地方自治体を事例として (The present conditions and problems of the demolition clearance policies of vacant houses). AIJ J Tech Design 18(39):709–714

Silverstein M, Cong Z, Li S (2006) Intergenerational transfers and living arrangements of older people in rural China: consequences for psychological well-being. J Gerontol Ser B Psychol Sci Soc Sci 61(5):S256–S266. Retrieved from http://www.ncbi.nlm.nih.gov/pubmed/16960239

Statistics Bureau (2017) Sekainotoukei 2017. Tokyo. Retrieved from http://www.stat.go.jp/data/sekai/pdf/2017al.pdf#page=16

Stevens C (2005) Measuring sustainable development. OECD statistics brief no. 10

Tacoli C (1998) Rural-urban interactions: a guide to the literature. Environ Urban 10(1):147–166. https://doi.org/10.1630/095624798101284356

Takegawa T (2010) 過疎農山村no okeru Koreisha no Seikatsu Jittai to Chiiki Fukushi no Kadai – Tottori-ken Nichinann-cho ni okeru Seikatsu Jittai Chosa Hokoku. (A study of living conditions of elderly people in depopulating rural districts and challenges of community care: a survey report on living conditions in Nichinan Town, Tottori Prefecture). Tottori Univ J Fac Reg Sci 7(1):2–22

Tamasato E (2009) Kourei Shakai to Nouson Kouzou [高齢社会と農村構造 Aging society and structure of agricultural community]. Showadou, Kyoto

Tsutsui K, Ebihara Y, Zushi N, Sakuma Y (2008) Survey report on a rural community development internship program in Kawamata Town, Fukushima Prefecture (in Japanese). Tottori Univ J Fac Reg Sci 5(2):85–96

Wilson GA (2008) From 'weak' to 'strong' multifunctionality: conceptualising farm-level multifunctional transitional pathways. J Rural Stud 24(3):367–383. https://doi.org/10.1016/j.jrurstud.2007.12.010

Wilson GA (2010) Multifunctional 'quality' and rural community resilience. T I Brit Geogr 35(3):364–381. https://doi.org/10.1111/j.1475-5661.2010.00391.x

Wilson GA (2012) Community resilience and environmental transitions, 1st edn. Routledge, New York

Yamamoto S, Nakazono M (2008) Tottori-ken Nishi-no-shima-cho no chuukounen settai iju sokushin jigyo ni yoru akiya katsuyou jirei – Nouson chiiki ni okeru akiya katsuyou shisutemu ni kansuru kenkyuu [鳥取県西ノ島町の中高年接待移住促進事業による空き家活用事例–農村地域における空き家活用システムに関する研究 Vacant house renovation by the migration promotion project for elderly in Nishinoshima-cho, Shimane Prefecture – study on the renovation system of vacant house in rural areas]. J Archit Plann 73(629):1485–1492

Chapter 6
Role in Framing in Sustainability Science — The Case of Minamata Disease

Motoharu Onuki

Abstract This chapter discusses multiple framings employed in Mainamata disease. Minamata disease is one of the major health problems caused by industrial pollution during Japan's high economic growth in the 1950s and 1960s. By conducting a historical review of Minamata disease, this chapter discusses typical framings applied in sustainability discourses in Japan, which have been led by pollution discourses. Two typical interpretations of Minamata disease are identified. One is that Minamata disease is a past event in Japanese history. It was a bitter experience, however thanks to this experience, the once-damaged Japanese environment became clean as environmental governance became stricter, regulations were established, and new environmental technologies were developed. Thus, one framing to Minamata disease is a historic event that Japan has learned lessons from the event, and something can be proud of how quickly Japan has recovered from such disaster. In contrast, even today, large-scale health examinations to understand the overall picture of methylmercury-derived health damage and to discover people with unrecognized symptoms continue. Therefore, Minamata disease remain unresolved and the local and national governments as well as Japanese society ignore the potentially hidden victims. The gap between these two framings is widening as the majority of the general public is unaware of the existence of the latter and some even believe that such humanitarian-conscious people are exaggerating their claims in an effort to obtain excessive compensation. To move forward, it is necessary to careful examine which part of framings people agree and disagree. By doing so, the essential nature of Minamata disease becomes clearer and collaboration among the people having different views may be possible. The ability to elicit and understand the true feelings of different stakeholders, the ability to apply different types of framings, and the ability to connect the people with different views, are critical when discussing a sustainability challenge that can be framed in diverse ways.

M. Onuki (✉)
Graduate Program in Sustainability Science - Global Leadership Initiative,
Graduate School of Frontier Sciences, The University of Tokyo, Tokyo, Japan
e-mail: onuki@k.u-tokyo.ac.jp

Keywords Minamata disease · Industrial pollution · Hidden victims · Historical review · Environmental governance

6.1 Introduction

This chapter discusses "framing" by using the Minamata disease as a case study. Minamata Disease is one of the major pollution diseases that Japan has experienced. Health problems – caused by industrial pollution during Japan's high economic growth in the 1950s and 1960s and that are treated as pollution diseases today – continue to plague the country. Since damage to the environment and people's health was severe, the overall impacts on Japanese society was considerable. These industrial pollution problems are exactly what gave birth to environmental engineering, environmental governance, environmental sociology, and many environmental-related academic disciplines including "environmental studies" in Japan as an integration of these disciplines. In the same vein, Japan's major emphasis on environmental aspects has led to the popularity of sustainability science in Japan as well. Thus, it follows that Minamata disease is one key factor leading to the origins of environmental studies and sustainability science in Japan. All the more because of this dubious history, I chose Minamata disease as a case study for discussing the issue of "framing" in sustainability science for this chapter.

In addition, Minamata disease not only represents part of the origin of sustainability science in Japan, but also represents one of the typical, ongoing current sustainability issues that involve several different framings even now. Although many people may be under the misconception that Minamata disease is an event of the past, several different ways of interpreting the Minamata disease case persist depending on different viewpoints and lingering disputes. When people encounter the term "lessons", they often feel that a misinterpreted nuance in that "lessons" generally implies that the matter in question no longer exists. However, the case in hand, Minamata disease, is yet to be resolved. To properly address the problems of Minamata disease, the skills and a sense of framing are necessary. This is another reason for choosing the Minamata Disease case as a case study.

This chapter explains the outline of Minamata disease case first, several framings of Minamata disease and then the importance of "framing" in sustainability science.

6.2 Overview of Minamata Disease

6.2.1 What is Minamata Disease?

Minamata disease is a disease of the central nervous system caused by eating seafood contaminated by methylmercury; in other words, a form of methylmercury poisoning. This was first officially acknowledged in May 1956, in Minamata City, located on the Kyushu island about 1000 km west from Tokyo; thus it came to be known as Minamata disease. Methylmercury damages specific parts of the central nervous system in the brain, each part with its own function. Depending on which functions become damaged, several types of symptoms appear: gait disturbance; loss of balance (ataxia); speech disturbance (dysarthria); muscle weakness; muscle cramps (disturbance of movement); decreased peripheral vision (constriction of visual fields); hardness of hearing (hearing disturbance); disturbances of sense of pain, touch or temperature (disturbance of sensation); and the inability to identify the form, size, weight, and texture of objects (stereo anesthesia, disturbance of sensation) by touch. In addition, another type of Minamata disease, *Congenital Minamata Disease*, is methylmercury poisoning of the fetus via the placenta caused when the mother consumes contaminated seafood during pregnancy. Such infant victims were born with a condition resembling infantile paralysis. (Minamata City 2007; George 2002; Harada 1995, 2004).

The methylmercury that caused Minamata disease was a product of a facility of the Minamata plant of Chisso Corporation that manufactured acetaldehyde, a raw material used in paint and plastic production. It was contained in the wastewater and discharged into the sea. Chisso Corporation, one of Japan's largest chemical companies at the time and still today, manufactured acetaldehyde at their plant in Minamata City in the 1950s. Unfortunately, however, Chisso unconscionably discarded the methylmercury that caused Minamata disease into the seawater. As a result, fish and shellfish in the sea became contaminated, and the people who ate them subsequently developed Minamata disease.

6.2.2 Delayed action

Minamata disease was caused by industrial pollution more than 60 years ago. However, even after official acknowledgement of the disease in 1956, Chisso Corporation continued unconscionable manufacturing of acetaldehyde for 12 more years. It was not until 1968 that the national government announced a consensus that the disease had definitely been caused by the methylmercury generated by Chisso Corporation.

During these 12 years, the spread of Minamata disease was left unabated. New victims emerged, and all victims have continued suffering not only from the disease itself but also from a social discrimination stigma. Several studies discussed why 12 years were required for proper action to be taken; but first, the rationale behind the failure to prevent the spread of the disease is attributed to the major impact the Chisso Minamata Plant had on the local economy. **The major portion of local taxes came from Chisso; and what is more, the mayor and many city council members were former Chisso employees [UN** Archives, 1992]. Also prevalent were concerns that any actions taken against the company and its Minamata plant would adversely affect Japan's strong economic growth at the time.

The second point concerns how to deal with scientific uncertainty. In 1959, the Minamata Disease Study Team of Kumamoto University's Faculty of Medicine reported that it had conclusive evidence that Minamata disease was caused by, methylmercury. However, scientists opposing this theory proposed other hypotheses, thus prolonging the scientific discussion and the search for commonly-agreed-upon causative substances. In addition, Chisso Hospital withheld announcing results of an experiment using a cat, although they confirmed the development of Minamata disease in the cat after feeding the cat the factory wastewater drainage. The government and manufacturing corporations have a moral responsibility to regulate and control pollution as soon as possible. However, Chisso's strong impact on the local and national economy which, under such circumstances, caused the government to be overly cautious for the wrong reasons and required more solid scientific evidence to develop an action plan. Science, on the other hand, always has its uncertainties. When a new hypothesis is proposed, scientists are compelled to validate it regardless of the time required and regardless of the urgency of needed attention victims. The government, unfortunately, delayed taking any constructive action for want of more conclusive results. In this way, government and scientists alike had no incentive to expedite the process. This "resonance between science and governance" is one cause of tardy for the delayed action (Shigeo Sugiyama, 2005).

6.2.3 Compensation and Relief for the Victims

The current scheme of compensating Minamata victims was established in 1973. This scheme requires that sufferers must be certified as a "Minamata disease patient" and approved by the governor of Kumamoto prefecture. A lump-sum conciliatory payment ranging from 16–18 million JPY was paid to these "certified patients" depending on the severity of the symptoms. To be certified, having a combination of several symptoms (disturbance in sensation and ataxia, etc.) is required; that is, those with only a single symptom remained uncertified.

More than 3000 sufferers have been "certified". However, depending on the level of methylmercury intake, more sufferers exist who have only one symptom such as disturbance of sensation or who have atypical symptoms. These types of sufferers have never been certified nor has any compensation ever been provided. Many have filed lawsuits petitioning to be certified, unfortunately with little or no success.

To resolve this situation, two political settlements were initiated in 1995 and 2010 to provide some relief to single-symptom sufferers. Even though such sufferers were not officially certified as "Minamata disease patients", they were recognized at least as "Minamata disease sufferers", and became able to receive some monetary relief in a lump-sum payment. About 11,500 sufferers received such relief in 1995 from the Japanese government, and some 65,000 sufferers applied for some form of settlement after the 1995 settlement.

The history of compensation and monetary and medical relief of Minamata disease is rife with the repetition of lawsuits and such settlements. Some lawsuits still continue by those who want to be certified, and others continue by those who will settle for some kind of monetary relief quickly. Even now, the exact number of total

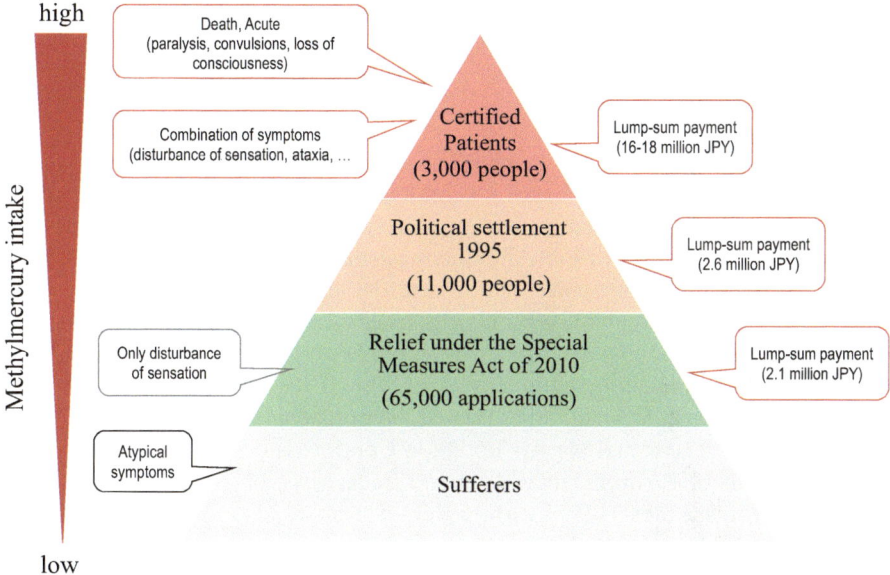

Fig. 6.1 Compensation and relief. (Adopted from Harada 2004). ∗Numbers have been rounded to thousands for easier understanding

victims suffering from disease. Methylmercury poisoning in the Minamata area remains unknown as shown in Fig. 6.1 (Harada 2004). Updating the knowledge on methylmercury poisoning must be continued.

6.2.4 How to Frame the Problems of the Minamata Disease

Based on the aforementioned circumstances, this section explains how people have interpreted the problems of Minamata Disease based on different framings.

6.2.5 What Was the Cause of Minamata Disease? (Scientific Framing)

The first way of framing the problems of Minamata disease focuses on what caused of Minamata disease. The answer to this question is clear now, although it took an inordinate amount of time (12 years) before the Japanese government officially declared that methylmercury is, with no doubt whatsoever, the substance acknowledged as the cause of the disease.

6.2.6   Why Did Environmental Governance and Pollution Control Technologies Fail? (Techno-Legislative FRAMING)

The second way of framing the problems of Minamata disease is how environmental governance and pollution control technologies worked. In reality, no environmental governance or regulation was in place when Minamata disease was first officially acknowledged in 1956. After many years, people finally realized the huge sacrifice – health damage to the victims and the social discrimination they were forced to endure – and public opinion supporting victims finally formed. Once this public opinion formed, the national government established environmental administration systems including enforcement of The Basic Law for Environmental Pollution Control in 1967 and other related environmental laws such as the Water Pollution Control Law and the Air Pollution Control Law in the following years, and foundation of Environmental Agency (predecessor of present-day Japanese Ministry of Environment) (1971). Moreover, enterprises also started following the new regulations designed to protect the environment or face consequences.

Industry and enterprises also started developing technologies to minimize pollution which led to rapidly and dramatically improved environmental quality. The technology and social system established in this way still functions well, achievements of which the Japanese can be proud; and what is more, of which developing countries might very well consider worth using as a role-model when confronting their own pollution and environmental protection issues.

6.2.7   How Much Does It Cost to Prevent or Recover from the Damage? (Economic Framing)

The third way of framing applies an economic viewpoint to Minamata disease. A study was done by a researcher group formed within the Japanese Environmental Agency, Japan (Study Group on Global Environment and Economy, 1991), in which the (estimated?) costs of preventing or recovering from the damage of Minamata disease were compared. The results indicated that the Costs of pollution control and the costs of prevention were significantly less than the costs of compensation for the caused damages and restoration of the polluted environment (see Table 6.1).

Table 6.1 Comparison of the cost of damage caused by Minamata disease in the area around Minamata Bay to the cost of pollution control and preventive measures

(million JPY per year)	
Cost for Pollution Control and Prevention Measures	123
Yearly average paid by Chisso Co.,Ltd., in the form of investments to control pollution	
Total damage amount	12,631
Health damage	7671
Yearly average of compensation benefits paid to patients under the Compensation Agreement	
Environmental pollution damage	4271
Yearly average amount of expenditure for dredging work in Minamata Bay	
Fishery damage	689
Compensation paid to the fishery industry computed as equal redemption of principal and interest prorated as yearly payment.	

Source: "Pollution in Japan – Our Tragic Experiences", ed. by Study Group for Global Environment and Economics, 1991

6.2.8   Were the 12 Years Required for Stopping the Acetaldehyde Process Long or Short? (Scientific Uncertainty Framing)

Another way of framing is considering whether the period of 12 years was long or short. The answer to this question is not simple. One view on this framing is "Not short, but there was no other way". As explained in the previous section, when neither environmental regulation nor previously available scientific knowledge was in place, "resonance between government and science" occurs (Shigeo Sugiyama, 2005). Taking 12 years was definitely a bitter experience, but valuable lessons were learned. First, science is a dynamic process. A certain level of uncertainty always remains, and science is always updating itself. Thus, government and society must not just wait for the "final conclusion" put forth by scientific study. Government must not use "remaining uncertainty" as an excuse for not taking any action. While science is ongoing research, society should prioritize human life, health, human rights, and the environment. Second, the polluter should bear the "costs of pollution prevention and control measures". This is known as the "polluter-pays principle" (OECD 1972). Thus, the company which has caused the pollution should pay for the recovery of the environment and compensation for the victims.

Another view is "12 years was long". Despite the lack of environmental governance and knowledge, the government should have reasonably been able to take some measures, at least measures to restrict the fish consumption soon after Minamata disease had been officially acknowledged. People of this framing are still fighting in court claiming that the government had the responsibility, and the authority, to prevent the spread of the disease in the early stage in the 1960s and 1970s.

6.2.9   Is the Mechanism of Minamata Disease (Methylmercury Poisoning) Fully Understood and are the Victims Properly for Damages? (Medical Framing and Its Social Implications)

Another way of framing is whether the entire mechanism of the disease was sufficiently understood from a medical point of view. Even though the substance causing the disease is has been clearly identified, the extent and condition of health damage of the methylmercury varies significantly according to the amount of exposure to methylmercury. However, as explained in the previous section, even today the relationship between the degree of exposure and the effect is not fully understood, except the cases of high dosages with acute and lethal effects. This is partly caused by insufficient data. Although mercury concentrations in hair and umbilical cords were used for estimating the level of methylmercury, collecting data from those whose symptoms were relatively milder was difficult. This is not only because such mild-symptom sufferers hesitated to at taking health examinations, but also because they had no inkling that they might have been affected by methylmercury in the first place. It has been pointed out that the certification criteria of Minamata disease victims tend to be limited to serious patients because this judgment resulted from political and administrative issues related compensation certification criteria. For this reason, scientific elucidation of how much damage has been caused by methylmercury poisoning became more problematic, and the total number of victims who suffered health damage by methylmercury, including those with relatively mild damage, remains elusive. Thus, a large-scale health investigation regarding the effects of methylmercury is has been carried out recently, accumulating more knowledge about the overall picture of mercury-derived health damage, and trying to "discover" people who suffered health damage by methylmercury, but who have neither come forward nor have become aware of personal damage.

However, compared to severe methylmercury poisoning symptoms found in the 1950s and 60s, the health damage found in today's health examinations is milder in comparison, and situation surrounding such health examinations differs greatly from those when the serious Minamata disease patients were discriminated against and persecuted in the early years in the 1960s and 1970s. In addition, when considering milder health damage (for example, numbness of hands), it becomes more difficult to distinguish such milder health damage from other health problems such as diabetes caused by non-methylmercury factors, which can also cause numbness in the hands, and becomes even more difficult to distinguish the impact from methylmercury derived from Chisso-polluted waters and other sources. Accumulating scientific knowledge is indispensable. It is also necessary to draw a line of administrative guidance on how far to compensate. It is important to "settle" at a certain level, rather than keeping "discovering hidden patients" forever, and to direct society towards looking ahead by promoting the rebuilding and redevelopment of divided local communities.

6.2.10   Who Must Decide the Criteria to Certify Victims for Compensation Before Relevant Scientific Knowledge is Sufficiently Accumulated, and How? (Time Consuming Nature of Science)

Time is needed to scientifically clarify something. Therefore, it is necessary to ask who must decide the criteria of for certifying victims for compensation before relevant scientific knowledge is sufficiently accumulated, and how. Some people believe that when problems occur, the cause must be scientifically identified first, and then actions must be taken to resolve problems by taking proper and sufficient measures based on scientific evidence. They expect that the certification criteria for victim compensation must be based on, and decisions made according to scientific evidence. Although some time is required, science must identify the cause and develop the countermeasures in the end. Accordingly, a certain level of risk must be tolerated for the sake of society's overall progress (including economic growth).

Others believe that science is always a dynamic process and that scientific clarification is time-consuming. Thus, such people think that society must decide the certification criteria from the viewpoint of other social values such as ethics and human rights, before science provides a clear knowledge base for humanitarian reasons. In the case of Minamata disease, scientific knowledge is important for explaining the cause(s) of the disease and for describing the extent of damage. However, science does not tell which level of damage must be compensated. These criteria have changed over time because people in society have gradually prioritized individual human rights and health more and more.

Society tends to wait for more convincing (conclusive) scientific evidence. From the perspective of the aforementioned first type of people who can tolerate the risk caused by insufficient scientific knowledge, the other type of people who prioritize human rights, and including the right to a safe environment conducive to good health, appear to incite society by raising unrealistic concepts, and "do not recognize how the society as well as its economy actually behaves. However, reality and truth are positioned just between the views of these two types. Thanks to the people with the more practical, humanitarian view, the number of victims finding relief has increased over the years. Thanks to the people with the more theoretical, society-first view, society has somehow managed to move forward.

6.2.11 Interpretations of the Problems of Minamata Disease and Importance of Understanding Framings Behind Them

Simply speaking, two typical interpretations of the problems of Minamata disease can be found nowadays. For most of the general public, the Minamata disease disaster is an event etched in Japanese history. This was a bitter experience. Thanks to this experience, however, the once-damaged Japanese environment became clean once again environmental governance became stricter, regulations were established, and new environmental technologies were developed. Thus, they believe that Japan has learned many lessons from these problems of Minamata disease, and can be proud of how quickly Japan has recovered from such disaster.

On the other hand, some people continue conducting large-scale health examinations on the effects of methylmercury for accumulating knowledge regarding the overall picture of methylmercury-derived health damage, and trying to "discover" people who are not yet known or recognized certified as victims by the government (or even not by themselves) because of their milder degree of damage. Because some people who have suffered from Minamata disease-like symptoms but are not yet been treated recognized as such, concluding that problems caused by Minamata disease have been resolved would be committing a most grievous error (Minamata City 2007). Some have taken their pleas to court to raise awareness of the government's responsibility to have taken measures to restrict fish consumption after Minamata disease was officially acknowledged. They also claimed that the local and national governments and Chisso Corporation tried to make the certification criteria for Minamata disease victims stricter so as to reduce compensation costs (Harada 2004). In their belief, the problems of Minamata disease most definitely remain unresolved, because uncompensated victims still remain and also because the local and national governments and society still ignore or pay insufficient attention to the potentially hidden victims.

The gap between these two typical interpretations is now widening. The majority of the general public having the first interpretation seems unaware of the existence of the latter interpretation. Several ongoing law cases and periodic Minamata disease-related health examinations seem to remain unnoticed by those waiting for scientific evidence to provide a conclusive solution, and putting humanitarian aspects as a lower priority. Some of them even believe that the latter humanitarian-conscious people are overexaggerating their claims in an effort to obtain excessive compensation (Higashijima 2010).

However, many interpretations are created from a combination of the several framings that I introduced in the previous section. Even if the interpretations completely differ from one another, all of them are apparently recognized as problems caused by Minamata disease. The difference is which part is being stressed as the crux of the problem, or what problems are identified. By carefully examining which framings people agree on and which framings people do not, the essential nature of Minamata disease should become clearer. And, this tends to lead to collaboration among the people having different views.

Important skills that are required when working with problems in the real world include the ability to elicit and understand the true feelings of different stakeholders and their views on a problem, the ability to be able to use different types of framings, and the ability to connect the people of different views. These are important skills regarding framing, especially for sustainability science.

Appendices

Appendix 1

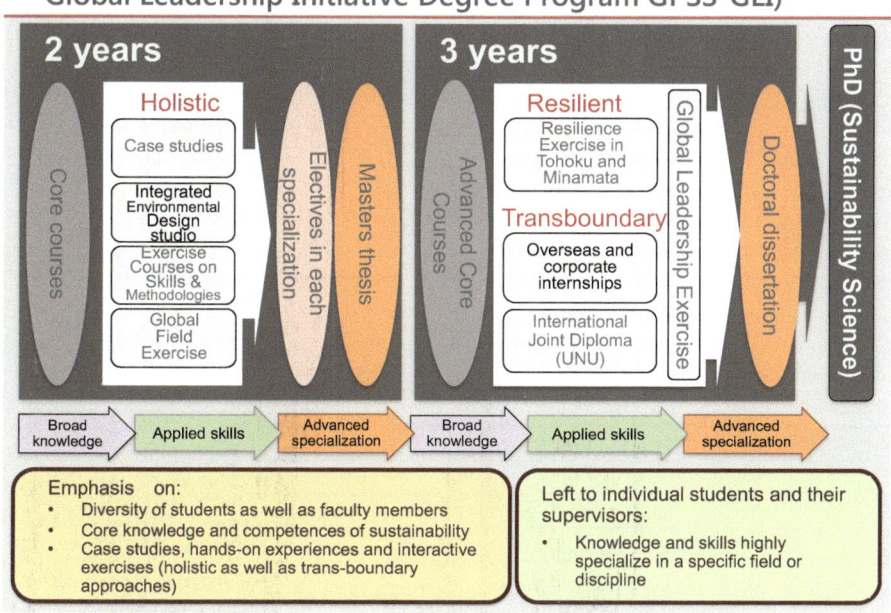

Appendix 2

	1st Year		**2nd Year**

	1st semester		3rd semester
Autumn	10 **Earth Systems Science MESB01**		4*7.5 **SELECTIVE COURSES (2 blocks, students choose 2 courses per block)**
	10 **Social Theory and Sustainability MESS32**		
	10 **Sustainability Science MESS33**		

	2nd semester		4th semester
Spring	7.5 **Governance of Sustainability MESS34**		30 **Master thesis MESM02**
	10 **Urban and Rural Systems and Sustainability MESS35**		
	7.5 **Economy and Sustainability MESS36**		
	5 **Knowledge to Action MESS37**		

Current 3rd semester selective courses (fall 2018)

1st block	MESS46	Social Movements
	MESS50	Landscape
	MESS51	Science and Politics of Climate Change
	MESS52	Global Health
	MESS56	Popular Culture

2nd block	MESS42	Water
	MESS47	Gender
	MESS53	Sustainability and Inner Transformation
	MESS54	Resilience and Sustainable Development
	MESS55	Political Ecology and Sustainability

References

George TS (2002) Minamata: Pollution and the Struggle for Democracy in Postwar Japan. (Harvard East Asian Monographs) Harvard University Press

Harada M (1995) Minamata disease: Methylmercury poisoning in Japan caused by environmental pollution. Crit Rev Toxicol 25(1):1–24

Harada M (2004) Minamata Disease (trans. By Tsushima S and George TS). Kumamoto Nichinichi Shinbun Culture & Information Center, translated from 「水俣病」原田正純 (1972) Iwanami Shinsho/Iwanami Shoten

Higashijima (2010) Naze Minamata-byo ha kaiketsu dekinainoka 「なぜ水俣病は解決できないのか?」 [Why can't we solve Minamata Disease Problems?] Genshobo弦書房

Minamata City (2007) Minamata disease: its history and lessons. http://www.minamata195651.jp/pdf/kyoukun_en/kyoukun_eng_all.pdf Accessed on 25 Aug 2018

Study Group on Global Environment and Economy (ed) (1991) Japan's experiences of environmental pollution – uneconomical economy that does not consider the environment. Godo Shuppan Publisher; 地球環境経済研究会(1991)"日本の公害経験―環境に配慮しない経済の不経済―"、合同出版

Sugiyama S (2005) *Minamata jirei ni okeru gyousei to kagakusha to media no sougo sayou*水俣病事例における行政と科学者とメディアの相互作用 [Interactions among administration, scientists, and the media in the case of Minamata Disease] in "Methodologies in Science, Technology, and Society", University of Tokyo Press, Yuko Fujigaki (ed) (in Japanese)

UN Archives (1992) Industrial pollution in Japan. ed. by Jun Ui http://archive.unu.edu/unupress/unupbooks/uu35ie/uu35ie0c.htm

Chapter 7
Time-Scale in Framing Disaster Risk Reduction in Sustainability

Miguel Esteban, Lilian Yamamoto, Lau Jamero, and Takashi Mino

Abstract Disaster Risk Reduction is one of the most important topics in sustainability science, seeking to examine the vulnerability and resilience of human life and society to natural hazards through the reduction and management of risks. However, disasters are caused by many different types of natural hazard events that take place in exposed and vulnerable areas across time spans. The size of the area and times-scale of the impact can also differ greatly. Possible actions to improve preparedness, countermeasures, actors or stakeholders involved, and person(s) in charge of these measures vary depending on the type of disaster. This chapter describes two different types of coastal issues, namely tsunamis and sea level rise, and the types of countermeasures available to either Japanese coastal towns or small coral islands. How these issues are perceived and dealt with will then be discussed from the point of view of time-scales, which affect the human perception of the problem.

Keywords Sustainability science · Risk management · Natural hazards · Tsunamis · Sea level rise

M. Esteban (✉)
Research Institute of Sustainable Future Society, Faculty of Civil
and Environmental Engineering, Waseda University, Tokyo, Japan

L. Yamamoto
South American Network for Environmental Migrations, São Paulo, Brazil

L. Jamero
Resilience Collaboratory, Manila Observatory, Manila, Philippines

T. Mino
Graduate Program in Sustainability Science-Global Leadership Initiative,
Graduate School of Frontier Sciences, The University of Tokyo, Kashiwa, Chiba, Japan

Department of Socio-Cultural Environmental Studies, Graduate School of Frontier Sciences,
The University of Tokyo, Kashiwa, Chiba, Japan
e-mail: mino@k.u-tokyo.ac.jp

© The Author(s) 2020
T. Mino, S. Kudo (eds.), *Framing in Sustainability Science*,
Science for Sustainable Societies, https://doi.org/10.1007/978-981-13-9061-6_7

7.1 Introduction

Natural hazards pose significant threats to the long-term sustainability of human settlements, as major events can overcome measures put in place to increase resilience, and can destroy the ability of socio-cultural systems to recover (Mino et al. 2016). Thus, how to prepare and manage natural disasters are two crucial topics for sustainability scientists, given that the very essence of sustainability science is to examine the long-term links between human life, well-being, and the environmental systems on which they are based. An event such as a landslide or an earthquake that occurs in a remote area and does not impact human life is considered merely a *natural event*, not a hazard or disaster. Such natural events are not included within the subject matters of sustainability science (at least in the narrow interpretation), and thus are excluded from consideration in this chapter.

When thinking about natural hazards, it is important to keep in mind that these events tend to repeat themselves at regular intervals, based on atmospheric or geological criteria that have a range of time spans (depending on the geographical location and nature of the hazard). The scale of the area and timelines involved can vary significantly, and humans (both individually and as a society) make conscious and unconscious calculations about such issues when designing socio-economic systems. The issue of hazard preparedness is thus clearly important, and the types of countermeasures, actors or stakeholders involved, and person(s) in charge of these measures will significantly differ depending on the type of disaster, level of development, and other characteristics of a given society.

For example, considering adaptation measures to the impacts of long-term climate change and "normal natural disasters" (which occur even without human-induced climate change) requires a different type of discourse; that is, they require a different "framing", especially concerning responsibility and the causes of the event (see for example Yamamoto and Esteban 2014). Other examples include building river dikes in preparation for the scale of heavy rains that may occur once in 50 years, constructing seawalls that anticipate a major tsunami that may occur once every 1000 years, and preparing for a volcanic eruption that can take place once in every 10,000 years and cover a huge area with its lava flow. These examples are framed differently and hence require different principles and processing for developing preventive measures and emergency plans.

Essentially, in the present chapter the authors argue that, when it comes to *large-scale* natural hazards, human societies tend to think in three different time scales (see Table 7.1). The first of these involves the largest scale event that is likely to take place during the life of one individual (i.e. individuals often think that they should prepare against it, as it is something that they can expect will happen during their own lifetime). The second relates to the largest scale event that can be thought possible in the course of that individual's civilization, and typically encompasses looking at time frames of hundreds to thousands of years. In this case the time scales used by different countries may differ significantly, depending on the length of their history and the quality of historical records and geological evidence. For example,

Table 7.1 Examples of how humans view different times scales related to two difference challenges

	Time Scales		
Phenomenon	One Human Lifespan (several decades to around one hundred years)	Human Civilization (hundreds to thousands of years)	Geological Time Scales (tens of thousands of years)
Tsunami (Japan)	**Level** 1 (around 10 m or less)	**Level 2** (over 10 m, to around 20 m or so)	**Level 3** (dozens of metres)
Sea level rise	Dozens of centimetres	A few metres	Dozens of metres

for the case of tsunamis, the Chilean society is currently looking at records of tsunamis from the arrival of the Spanish in the sixteenth century (Aranguiz 2015), the Japanese are usually going back until the seventh century (San Carlos et al. 2017), and the Greeks are attempting to gain insights from as far back as the end of the Minoan Civilization (2000–1400 BCE) and the volcanic eruption of Santorini (circa 1646 BCE) (Pareschi et al. 2006). Essentially, individuals currently alive think about the consequences that these events will have on their descendants and on the long-term survival of their culture and traditions, something that tends to weigh quite heavily on the cultural conscience of many humans.

Finally, geological time scales are typically outside the calculations of even the more advanced societies, and represent acts of society-level *force majeure* from which it is thought impossible to protect or adapt, at least under present technology levels. These are often disregarded by individuals, given the difficulty in relating to the time scales involved.

The discussion of time scales in implementing effective disaster risk reduction is particularly important, although not explicitly stated, in the most recent global disaster risk reduction framework – the Sendai Framework for Disaster Risk Reduction 2015–2030. The framework's goal is to "*prevent new and reduce existing disaster risk through the implementation of integrated and inclusive economic, structural, legal, social, health, cultural, educational, environmental, technological, political and institutional measures that prevent and reduce hazard exposure and vulnerability to disaster, increase preparedness for response and recovery, and thus strengthen resilience*" (UNISDR 2015). The framework promotes that societies should understand their risks and then plan and act accordingly, in order to absorb known and unknown shocks and disturbances. This involves understanding the retrospective and prospective intricacies of risk across time scales.

The framework has four priorities for action: (1) understanding disaster risk, (2) strengthening disaster risk governance to manage disaster risk, (3) investing in disaster risk reduction for resilience, and (4) enhancing disaster preparedness for effective response and to "Build Back Better" in recovery, rehabilitation, and reconstruction (UNISDR 2015). All four priorities are to be implemented at the local, national, and global scales with emphasis on promoting long-term resilience towards disasters. The time scales in which disasters could occur enable the proper framing of preventative actions and strategies in implementing these priorities for action.

In the present chapter we discuss how these different time scales impact the framing of disaster preparedness and risk reduction using two different case studies. The first relates to tsunami disaster risk management in Japan, and the second to sea level rise using the point of view of low-lying islands and coral reefs as an indication of how time scales affect the perception of the problems involved.

7.2   Natural Hazard Return Periods: Tsunami Classification in Japan

On March 11, 2011 a large earthquake of magnitude 9.0 on the Richter scale occurred off the northeast coast of Japan, generating a devastating tsunami that inundated over 400 km^2 of land, and caused large numbers of casualties (Mori et al. 2012, Mikami et al. 2012; Ogasawara et al. 2012). Along the Sendai plain in northern Tohoku the maximum inundation height was 19.5 m, and the tsunami propagated as a bore around 4–5 km inland, with maximum run-up-heights of 40.4 m being measured (Mori et al. 2012). Widespread devastation ensued, as the waves engulfed entire settlements, with everything but the sturdiest of buildings being completely washed away (see Fig. 7.1). This event is now known as the *2011 Tohoku Earthquake Tsunami*.

This event has transformed the way that the Japanese engineering and coastal zone management community think about tsunamis. Their approach to time scales is discussed in detail in the rest of this section.

7.2.1   History of Tsunamis in Northern Japan (Tohoku region)

Historically, the *2011 Tohoku Earthquake Tsunami* was one of the worst tsunamis that has affected Japan since records began. The Sanriku coastline, which extends northwards from the city of Sendai, has been frequently affected by tsunamis. The recorded history in the region goes back over 1000 years, and five major destructive tsunamis are all well documented (Watanabe 1985):

1. Jogan (869),
2. Keicho (1611),
3. Meiji-Sanriku (1896),
4. Showa-Sanriku (1933), and
5. Chile (1960).

In fact, the *2011 Tohoku Earthquake Tsunami* has been described as a one-in-a-1000-year event, resembling the *Jogan Tsunami* in A.D. 869 (Sawai et al. 2006). The description of this *Jogan Tsunami* actually appears in a historical document known as the *Sandai-Jitsuroku*, which documents how the wave flooded a wide

Fig. 7.1 Coastal settlement of Arahama, formerly located in the vicinity of Sendai City, Japan. All residential buildings in the town were destroyed by the power of the tsunami

coastal area of Tohoku (the northern region of Japan that encompasses the Sanriku coastline and the Sendai plains, amongst other areas), killing some 1000 people (though population density at the time was significantly lower than at present). There are no records concerning the *Jogan Tsunami*, though some tsunami deposits found in sediment layers in the Sendai Plain, as well as along the Sanriku Coast, have allowed researchers to identify the area which was likely to have been inundated by this event (Minoura et al. 2001).

Since the Edo Era (1603~1867), the number of written records increased substantially, and thus tsunamis have been better documented. The Keicho *Tsunami* (1611), which attacked a wide coastal area from Hokkaido to Sanriku, was one of the most destructive tsunamis in this period, and in the Tohoku regions waves travelled up to 4 km inland (Sawai et al. 2006). Since the beginning of industrialization during the Meiji Era (1868–1912), the Sanriku coast has experienced three major tsunamis. The first of these three tsunamis is known as the *Meiji-Sanriku Tsunami*, which caused some 22,000 casualties. Although the magnitude of the generating earthquake was comparatively modest, the maximum tsunami height reached as high as 20 m. The second tsunami is referred to as the *Showa-Sanriku Tsunami*, which caused around 3000 casualties along the Sanriku coastline. Finally, the 1960 *Chile Tsunami*, which was triggered by an earthquake of magnitude 9.5 on the Richter scale in Chile, reached the Sanriku coastline and caused over 100 casualties.

7.2.2   New Tsunami Classification System

The *2011 Tohoku Earthquake Tsunami* led to major discussions within the Japanese coastal engineering and management community about whether hard measures (such as breakwaters or dikes) are preferable over soft measures (such as tsunami warning systems and evacuation plans) to protect the coastline and the communities situated next to it (Shibayama et al. 2013). Eventually, the concept of Level 1 and Level 2 tsunamis emerged based on ideas concerning time scales and the likelihood of an area being affected by such events. These concepts are widely used today, and formed the cornerstone of reconstruction philosophy in the aftermath of the event (as will be expounded upon later in this chapter). It is important to note that this classification is based on the frequency of these events, and that the exact period of return of each of the events has yet to be fixed, though there is a clear feeling that one of them relates to human life-spans, and the second to civilization time scales. The two levels would be:

Level 1 Tsunami Events with a return period of several decades to 100+ years (essentially, the Japanese expression which has been used in coastal engineering discussions would translate as a return period from 50–60 to 150–160 years). Although the height of the wave would depend on the event and location, Level 1 Tsunamis refer to waves which are *comparatively* low in height, typically less than 7–10 m.

Level 2 Tsunami These events have return periods of between one hundred to a few thousand years. The tsunami heights are much higher, and encompass waves over 10 m in height, and sometimes even up to 20–30 m. Clearly, both the 2004 *Indian Ocean* and 2011 *Tohoku Earthquake Tsunamis* fall under this category.

It is important to note that given the nature of the propagation process, a given event might represent a Level 2 tsunami for a certain area or country, yet only a Level 1 event for other places. For instance, the 2011 *Tohoku Earthquake Tsunami* was clearly a Level 2 event in northern Japan, though by the time it reached Chile the tsunami was only a Level 1 event.

Events such as meteorite impacts or underwater landslides, which can cause devastating waves over 50 m in height, are outside the scope of this classification. These types of hazards would have return periods of tens or hundreds of thousands of years, and would completely devastate a coastline, reaching dozens of kilometres inland and probably rendering useless any evacuation strategy in place. One could actually talk about "Level 3" events, and it is unclear whether present day technology is advanced enough to protect human society from them. To the authors' knowledge, no strategies are currently in place anywhere in Japan (arguably the country in the world which has invested the most (Mori et al. 2012) in improving resilience against natural disasters in general, and tsunamis in particular) or any other country to protect against such events. This highlights how the time scales involved in an event dictate the type of actions (or complete lack of actions, for events with very high return periods) employed to improve resilience and mitigate the consequences of a given hazard.

7.2.3 Implications of Time Scales on Measures to Improve Resilience

The debate on whether hard or soft measures are better suited to protect against coastal hazards has used the concepts of Level 1 and Level 2 tsunamis to understand the role that each type of countermeasure has on improving resilience. At present, the idea that hard measures alone can always protect against the loss of life is no longer accepted; instead, it is thought that coastal structures should play a role in attempting to protect property against Level 1 events. Thus, the function of safe-guarding human life should fall onto soft measures, which should be designed against Level 2 tsunamis. Nevertheless, hard measures might aid evacuation, and their influence can be considered when thinking about the design of evacuation systems. For example, an assessment of the effectiveness of Kamaishi Bay mouth breakwater shows that the structure could have contributed to reducing inundation heights by around 40 to 50%, and could have provided extra time for local residents to evacuate (data from the Tohoku Earthquake Tsunami Joint Survey Group and PARI 2011, also see Shibayama et al. 2013).

However, it is important to note that the cost of using hard measures for tsunami protection is often rather high, and their effectiveness against Level 2 tsunamis is unclear, though a number of lessons have been learnt after the 2011 event (Jayaratne et al. 2016). It is also important to consider whether coastal areas are a place for recreation, or the source of potential threats. Japan is a country that regularly experiences many different types of coastal natural disasters, and countermeasures against typhoons and tsunamis require the construction of coastal defences, river embankments and other engineering structures (such as landslide countermeasures, which can take place because of high precipitation (perhaps even 150 mm of rain in 1 h) during the passage of a typhoon). Thus, important decisions must clearly be taken by society about which areas should be designated as residential areas, how those areas should be protected, and the consequences to the rest of the country if one of these areas suffers from a natural disaster (Table 7.2).

7.2.4 Case Study: Otsuchi Town

In order to illustrate reconstruction patterns and how this classification of Level 1 and Level 2 tsunamis affects the way of thinking about future hazards, it is worth looking at a case study of one city in Japan. The town of Otsuchi was particularly devastated by the 2011 event, with recorded inundation heights of 10–14 m and run-ups of around 25 m (Mori et al. 2012). The initial wave arrived just 34 min after the earthquake (Yamao et al. 2015), which explains the large numbers of casualties and the challenge that it represented from the long-term demographic sustainability of the town. Prior to the 2011 event, Otsuchi had a nominal population of around 16,000 people, and out of these 803 people died, 431 are still missing, and a further

Table 7.2 Summary of the philosophy regarding the use of hard and soft measures to protect against Level 1 and Level 2 tsunamis in Japan

Tsunami level	Hard measures	Soft Measures
1	*Primary function* Protect property *Secondary function* Help in the protection of lives	*Primary function* Protect lives Tsunami early warning and evacuation system
2	*Primary function* Possibly provide residents with some extra time to evacuate area	*Primary function* Protect lives
	Generally ineffective	Tsunami early warning and evacuation system

50 lost their lives because of the indirect consequences of the tsunami (e.g. in the aftermath of the disaster some people died because they lost access to medicines needed to treat chronic illnesses (Esteban et al. 2015). Regarding the damage, 3359 buildings were completely destroyed and another 713 suffered major or partial damage (Esteban et al. 2015).

It is worth noting that the town has a long history, and thus prior tsunamis have been well documented, as it served as a provincial capital during the Edo era. By 1948 the central downtown area was concentrated on the side of one of the hills, with the areas close to the sea left undeveloped, given that they were destroyed by previous tsunamis such as the 1896 *Meiji-Sanriku* and 1933 *Showa-Sanriku* events, and thus local inhabitants had a strong cultural memory of such types of disasters (Esteban et al. 2015; Esteban et al. 2013). Nowadays, economic activities are based around the service sector, with a significant contribution of salmon fishing, aquaculture of scallops and seaweed, and the fish processing industry to the local economy.

Local authorities are aware that the tsunami walls protecting the town were unsuccessful in stopping the tsunami wave, and that the only inhabitants that survived were those that evacuated or were in areas that were not at risk. Thus, this classification of tsunami levels and the need to further emphasize evacuation were accepted, as the ultimate objective should always be to preserve lives (Esteban et al. 2015). Nevertheless, the types of interventions that are currently being considered can be classified into three different layers of protection:

Layer 1-Prevention: consists of breakwaters or dykes aimed at preventing seawater from inundating the land;

Layer 2-Spatial Solutions: involves spatial planning and adapting buildings to mitigate losses if flooding does take place, and includes relocating important buildings to higher ground (which essentially means that areas closer to the sea should be considered as sacrificial, and only dedicated to industrial buildings or parks)

Layer 3-Emergency Management: involves the use of disaster plans, risk maps, early-warning systems, evacuation, and medical help, and mainly focuses on measures that reduce risks to human life.

Essentially, the idea is to move towards a more resilient and flexible system that relies on multi-layer (Tsimopoulou et al. 2012, 2013).

Given financial consideration and guidance from the national government, following the concepts of Level 1 and Level 2 tsunamis described earlier requires that Layer 1 coastal defences should be rebuilt against a Level 1 tsunami, but not necessarily against a Level 2 tsunami. Prior to the 2011 event, the highest tsunami walls in the town were built up to a height of +6.4 m T.P.[1]

Simulations carried out by the national and prefectural governments indicate that the *meiji-sanriku* tsunami should become the benchmark for a level 1 event (which flooded otsuchi to a level of around +11.5 m t.p.). Nevertheless, because the town is located close to Kamaishi city it was decided that most of the tsunami walls would be built to the same inundation height as that expected in Kamaishi, i.e. to a level of +14.5 m t.p. Simulations of the 2011 event indicate that even for such a wall partial overtopping is possible, allowing some water to flood the land behind it. Thus, Layer 2 countermeasures, in the form of "land adjustment," are also necessary. Under the new land use maps. The areas immediately adjacent to these walls can only be utilized for fishing industries and parks.

However, in some neighbourhoods in Otsuchi, local residents decided that the wall should be rebuilt to the same height as that which existed prior to the tsunami (+6.4 m T.P.), much lower than the walls protecting the main downtown area, in some cases for aesthetic reasons (Esteban et al. 2015). It is important to note that such considerations can be quite important for the mental well-being of the population, as the connection to nature and the sea is usually very important for communities living close to the coastline. Also, the economy of the area has a significant component of tourism, and thus in many cases thinking about how to preserve the natural beauty of the land is also necessary. Nevertheless, in order to compensate for this, Layer 2 countermeasures have been significantly improved by raising the entire residential areas by over 8 m, to bring them to a height of almost 15 m above sea level, which is higher than the inundation height during the 2011 tsunami. Aside from these special areas, the entire central part of Otsuchi has been elevated by at least 3 m (see Fig. 7.2). This would mean that, in combination with the Layer 1 measures described earlier, the town would not be flooded if an event like the 2011 event were to recur. Paradoxically, this seems to imply that the defences have been planned against a Level 2 tsunami, which poses significant questions regarding the long-term sustainability of the Japanese country as a whole (Can the government of the country afford to financially protect all communities in Japan to the same degree?). However, considering such question is outside the scope of the present chapter.

---

[1] These heights are presented relative to Tokyo Peil (T.P. corresponds to mean sea level of Tokyo Bay).

Fig. 7.2 The entire downtown area of Otsuchi is being elevated, as part of the strengthening of Layer 2 countermeasures

7.3   Sea Level Rise and Low-lying Lands

Climate change and sea level rise are expected to pose considerable challenges to human civilization in the coming centuries, and their consequences in the course of the twenty-first century alone are widely discussed in media and academia. However, the problem posed by sea level rise depends on the time scales on which it is considered. Though present-day society concerns relate to the likely sea levels within the lifespan of those alive today, it is important to remember that such changes will not stop by the year 2100. Hence, when considering the sustainability implications of sea level rise, it is important to keep this factor clearly in mind, as it will influence how to deal with the issues involved. To illustrate how time scales influence the choice of adaptation strategies, the authors will describe the problems being faced by low-lying coral islands, which traditionally depend on coral reefs to supply the materials necessary to compensate for sea level rise (Yamamoto and Esteban 2014).

7.3.1 Past Sea Level Rise and Twenty-First Century Projections

The International Panel on Climate Change fifth Assessment Report (IPCC 5AR) mentions how surface temperatures have oscillated for millions of years following glacial cycles. This in turn has influenced sea levels, which have risen and fallen according to such variations in temperature (because of thermal expansion of the oceans and the melting or accumulation of water in polar caps). During most of the twentieth century the global mean sea level rose by around 1.7 mm per year on average, though this intensified to 3 mm per year towards the end of the century (IPCC 5AR). The IPCC 5AR estimates that sea levels could rise by between 26 and 82 cm by 2100, substantially higher than the 18–59 cm projections that had been given by IPCC 4AR. So-called "semi-empirical methods" (see IPCC 5AR) such as those by Vermeer and Rahmstorf (2009) provide more onerous predictions, indicating sea level rise for 1990–2100 could be in the 0.75–1.9 m range.

There is little doubt that climate change – and consequently sea level rise, as it is greatly affected by global temperatures – is mostly being driven by the release of greenhouse gases into the atmosphere. Current world efforts to reduce greenhouse gas emissions, centred around the United Nations Framework Convention for Climate Change (UNFCCC), have not yet convincingly managed to halt their increase, despite the signing of the Paris Agreement in 2015. However, even if emissions were to reduce, the IPCC 4AR points how "if actions are taken to reduce the emissions, the fate of the trace gas concentrations will depend on the relative changes not only of emissions but also of its removal processes" (Bindoff et al. 2007). This means that it could potentially take a very long time for the Earth to revert to its current condition. As CO_2 emissions continue unabated, global temperatures will inevitably continue to rise unless drastic action is taken to curtail them. Such effects can very well lead to the flooding of low-lying deltaic areas such as the Mekong delta (see Nguyen et al. 2013; Takagi et al. 2014; Nobuoka and Murakami 2011) or atoll islands (Yamamoto and Esteban 2014), unless significant adaptation measures are implemented.

The IPCC 5AR discusses the long-term climate change and commitment up to the year 2500. Essentially, if greenhouse gas concentrations rise to between 500 and 700 ppm CO_2, sea level rise could exceed 1.5 m by the year 2300; or if concentrations were to exceed 700 ppm CO_2 sea level rise could surpass 3 m by 2300, reaching almost 7 m by 2500. Essentially, the most optimistic scenarios related to sea level rise require a positive outcome of UNFCCC efforts and negotiations. The entire Earth climate system, however, exhibits a certain lag, due to the thermal inertia of the oceans. The oceans will gradually absorb heat from the atmosphere, and this will lead to the heating of the top layers, gradually extending deeper into the ocean. Even when air temperatures stop increasing, the heat absorbed by the oceans will be slowly released, meaning that the oceans would become a "very weak heat source" and dampen the decline of surface atmospheric temperatures (Schewe et al. 2011). Hence, the effects of current increases in CO_2 concentrations will manifest themselves in the future, much in the same way that present climate change is being caused by past CO_2 emissions. Also, the process of CO_2 removal from the atmosphere is quite complex, and although more than half of the CO_2 emitted is removed within a century, a fraction remains in the atmosphere for millennia. Another mechanism that slows down global cooling is the change in oceanic convection, which enhances ocean heat loss in high latitudes and reduces the surface cooling rate by almost 50% (Schewe et al. 2011). In this sense, greenhouse gases released at present "commit" us to certain future effects, as yet unfelt.

Simulations by Schewe et al. (2011) suggest that if a maximum warming of 1.5 °C is reached by middle of the twenty-first century (for CO_2 concentrations of just over 550 ppm), then temperatures will likely decline slowly afterwards and reach present-day levels by 2500. This would be achieved by GHG concentrations peaking in 2040 and declining subsequently to become negative after 2070. If this is achieved, the rate of sea level rise caused by thermal expansion (where the volume of the seawater would increase due to the change in temperature) would continue for over 200 years after the peak in air temperatures and stabilize around 2250. The rate of temperature decrease is significantly slower than the current rates of temperature rise, and are on average around −0.16 °C per century. This slow rate of cooling is quite significant, and highlights the need to rapidly reduce emissions of greenhouse gases and the importance of current climate negotiations between different countries. Not achieving these objectives could result in global warming continuing for much more prolonged periods of time. Another scenario by Schewe et al. (2011) shows how CO_2 concentrations of almost 1500 ppm by 2100 can see warming of up to 8.5 °C and result in 1.3 m of sea level rise due to thermal expansion alone by 2250 and a 2 m rise by 2500 (though some of the ranges given in the IPCC 5AR are much higher, as noted earlier).

Past geological records of sea level rise indicate that sea levels could very well have been much higher than current levels. For example, during a period known as the marine isotope stage 11 (MIS 11, 401 to 411 ka), global temperatures may have been 1.5–2.0 °C higher than those on the planet today, with sea levels possibly also being 6–15 m higher (IPCC 5AR). Also, during the last interglacial period, temperatures might have been 1–2 °C higher than pre-industrial levels, with sea levels several metres (around 4–8 m, see IPCC 5AR) higher than at present. Since then, in the late Holocene (some 12,000 years ago) it is likely that global sea levels rose 2 to 3 m to near present-day levels. All this indicates the necessity to factor time scales into the framing of the problems that coastal areas face because of sea level rise.

Finally, it is worth pointing out that these projections have only been made at the global level and on general, average terms. Although islands are expected to be most vulnerable to sea level rise, precise information about how much sea level rise a particular island or island state will experience are yet unavailable. Due to this, it is important to further discuss the particular impacts of sea-level rise on coral islands and the communities that inhabit them.

7.3.2   Island Communities

While environmental factors such as the natural survival of islands are important in predicting the sustainability of the communities living in them, recent studies about the impacts of sea-level rise have highlighting social factors, such as human adaptation, as a greater determinant (Perch-Nielsen et al. 2008; Gibbons and Nicholls 2005). However, given the lack of climate projections at the local level, time-scales are also important in understanding the impacts of climate change and the ways that communities can adapt to them. In particular, when considering the survival of island communities, differentiating between short-term (decades) and long-term (centuries) impacts, as well as potential short-term and long-term adaptation strategies is critical (Fig. 7.3).

In the long-term, mass migration theory suggests that, because of land loss, entire populations could be driven out of their homes (Yamamoto and Esteban 2014). However, in the short-term, the theory also argues that, due to disruption in food and water supply through saltwater intrusion, mass migration could also happen well before total land loss (Keener et al. 2012). Although this theory is dominant in the discussion surrounding sea-level rise adaptation, historical and empirical evidence so far indicates otherwise.

Fig. 7.3 Low-lying coral islands are particularly vulnerable to the consequences of sea level rise, unless coral species can successfully adapt to changing ocean conditions

Fig. 7.4 Adaptation strategies to rapid relative sea level rise in small islands in the Philippines, caused by earthquake induced land subsidence (see Jamero et al. 2016)

Historically, the world's biggest coastal cities have been able to prevent tidal flooding through engineering methods and land use planning (Nicholls and Cazenave 2010). A case study from the central Philippines also provides evidence that it is indeed possible for island communities to adapt to changes in sea levels of less than 1 metre, even if they happen quickly as a consequence of earthquake induced land subsidence (Jamero et al. 2016). Essentially, the 2013 Bohol Earthquake in the Philippines caused a number of small islands to immediately subside by up to 1 metre, which means that they are now flooded during high tides. The strategies implemented by these communities, mainly on a self-funded basis, are designed to accommodate the effects of tidal flooding, including building stilted houses (see Fig. 7.4) and raising the roads and floors of important community buildings (such as schools and chapels). The communities have also changed their evacuation behaviour to protect lives from the potential risks of passing typhoons (including high waves and storm surges; see also Jamero et al. 2017).

However, pressure from coastal settlements and bleaching events due to high ocean water temperatures are causing large-scale damage to many coral systems around the planet. Essentially, from the point of view of a human's lifespan, large-scale mortality and devastation will take place in these fragile ecosystems in the coming decades. Biologists fear that thermal stresses caused by sea water warming and ocean acidification – brought about by the absorption of CO_2, which reduces the ability of corals to produce their skeletons – will result in more prolonged episodes of bleaching and increased mortality. Veron et al. (2009) estimate that by the 2030s coral reefs could very well be in severe danger throughout the world, which could create many problems for small island nations, given the reduction in sediment supply that this would represent. Given that these islands are geomorphologically very dynamic, it is unlikely that they would disappear (Kench et al. 2009, Webb and Kench 2010), though this could lead to greater damage due to high waves and the need for substantial adaptation strategies. Nevertheless, evidence proves that it is indeed possible for coastal communities to adapt to changes in sea levels of less than 1 m, even if those changes happen quickly (Jamero et al. 2016).

However, when looking at a time scale of several centuries, the situation changes. Sea level rise will not stop by the year 2100, and it would be increasingly difficult for the inhabitants of small islands to adapt to changing water levels if all the coral dies. This could eventually lead to societal collapse in the islands, forcing their inhabitants to migrate (Yamamoto and Esteban 2014, 2016).

When looking at a longer time frame, it is clearly possible that coral communities will somehow adapt to these new changes (through the recruitment of new, better-adapted species, Kench et al. 2009), and there is evidence that coral reefs have adapted in the past to changing conditions (Kench et al. 2009). The Census of Marine Life (2010) found relics of cold water corals off Africa's Mauritanian coast extending over 400 km in waters 500 m deep in one of the world's longest reefs. This highlights how corals have continuously evolved to adapt to changing ocean conditions, and this will likely continue to happen in the future, though the time frames involved are unclear. From an evolutionary point of view, coral diversification has occurred in pulses, and mass extinctions have caused bottlenecks in the evolution of corals (Simpson et al. 2011). It is thus possible that major increases in coral mortality also retard the time it takes the species to re-adapt to the new environmental conditions, as a less diverse coral population has less of a genetic base from which to re-adapt. However, such evolution may already be taking place, and some evidence indicates that certain species around the Persian/Arab Gulf may have adapted or evolved to withstand higher sea temperatures, perhaps as much as 35 °C, that would normally prove fatal to corals elsewhere (Hume et al. 2016).

Researchers in Japan have also stated that there is large scale evidence that several major coral species have begun spreading polewards at speeds of up to 14 km/yr. (Yamano et al. 2011), showing how species can also adapt by moving to other areas of the planet where they find more favourable conditions. Thus, these areas may serve as a refuge for tropical corals in an era of global warming and could later move towards the equator again if and when temperatures return to their present values. One absolute limiting condition to this shifting of species towards the poles may be acidity, as corals stop growing in pH concentrations of 7.7 or lower (Fabricious et al. 2011), although it appears that some coral species can survive in conditions of higher acidity. It thus appears that corals could somehow adapt to a changing environment (by evolutionary or migration modes), assuming that this lower level of ocean acidification is not reached. This adaptation happens even with gradually rising water levels, and there is some evidence that the Maldives may have originated during a period of sea level rise, as noted by Kench et al. (2005).

Thus, it is clear that a variety of time scales and mechanisms must be taken into account when considering the long-term sustainability of small low-lying island nations. Currently, scientific knowledge regarding what is mostly likely to happen is incomplete, and much depends on what happens to coral reefs along different time scales, which in turn will determine the types of adaptation strategies that are required. If corals adapt within decades, the morphological resilience of the islands may prove sufficient to avert disaster, as their inhabitants can resort to rising the islands using dredged materials or coral stones (providing that the biological limits of sediment supply are not breached). When considering longer time scales, it is more difficult to see how a good solution can be reached unless corals adapt, though a number of options could be available.

7.4   Conclusions

In the present chapter the authors have attempted to show how the time scale through which a sustainability scientist looks at a problem will clearly influence disaster preparedness and management. Looking at these issues is indispensable for examining the sustainability of human life and well-being in the face of natural hazards, and for attempting to develop measures to improve the resilience of human societies. Such kind of measures should always carefully consider the different stakeholders and actors involved in order to attempt to arrive at a holistic assessment of the problem, taking into account present conditions and how things are likely to change in the middle to long term (hence once again the importance of looking at different time scales).

To illustrate the issues involved, the chapter presented two different types of hazards, namely that of tsunamis and the problems that will be brought about by sea level rise and ocean acidification (and the consequences it will have on coral reefs). Then, the case study of the reconstruction of Otsuchi town in northeastern Japan after the 2011 Tohoku Earthquake Tsunami and small islands of the coast of Bohol following the 2013 Bohol Earthquake in the Philippines, were presented. The discussion of both cases proved that the solutions to any given problem depend on the time scale considered, though much uncertainty exists about the likely future consequences and return periods of any given hazard. This highlights the need for more research on the topic of natural hazards taking into account long-term sustainability, in order to attempt to improve resilience and minimize the consequences of such events.

Acknowledgements A part of the present work was performed as a part of activities of Research Institute of Sustainable Future Society, Waseda Research Institute for Science and Engineering, Waseda University.

References

Aranguiz R (2015) Tsunami resonance in the bay of concepcion (Chile) and the effect of future events. In: Esteban M, Takagi H, Shibayama T (eds) Handbook of coastal disaster mitigation for engineers and planners. Butterworth-Heinemann (Elsevier), Oxford

Bindoff NL et al (2007) Climate change 2007: the physical science basis. Contribution of working group I to the 4th assessment report of the intergovernmental panel on climate change. Cambridge University Press, New York

Census of Marine Life (2010) First Census of Marine Life 2010, Highlights of a Decade of Discovery wwwcomlorg Accessed 03 August 2011

Esteban M, Tsimopoulou V, Mikami T, Yun NY, Suppasri A, Shibayama T (2013) Recent tsunami events and preparedness: development of Tsunami awareness in Indonesia, Chile Japan. Int J Disast Risk Re 5:84–97

Esteban M, Onuki M, Ikeda I, Akiyama T (2015) Reconstruction following the 2011 Tohoku Earthquake Tsunami: case study of Otsuchi Town in Iwate prefecture, Japan. In: Esteban M, Takagi H, Shibayama T (eds) Handbook of coastal disaster mitigation for engineers and planners. Butterworth-Heinemann (Elsevier), Oxford

Fabricius KE, Langdon C, Uthicke S, Humphrey C, Noonan S, De'ath G, Okazaki R, Muehllehner N, Glas MS, Lough JM (2011) Losers and winners in coral reefs acclimatized to elevated carbon dioxide concentrations. Nat Clim Chang 1:165–169

Gibbons SJA, Nicholls RJ (2005) Island abandonment and sea-level rise: an historical analog from the Chesapeake Bay, USA. Global Environ Chang 16(1):40–47

Hume BCC, Voolstra CR, Arif C, D'Angelo C, Burt JA, Eyal G, Loya Y, Wiedenmann J (2016) Ancestral genetic diversity associated with the rapid spread of stress-tolerant coral symbionts in response to Holocene climate change. PNAS 113(16):4416–4421

IPCC (2007) IPCC fourth assessment report: climate change 2007: the physical science basis [IPCC 4AR]. In: Solomon S, Qin D, Manning M, Chen Z, Marquis M, Averyt KB, MMB T, Miller HL (eds) Contribution of working group I to the fourth assessment report of the intergovernmental panel on climate change. Cambridge University Press, Cambridge, UK/New York

IPCC (2013) IPCC fifth assessment report: climate change 2013: the physical science basis [IPCC 5AR]. In: Stocker TF, Qin D, Plattner G-K, MMB T, Allen SK, Boschung J, Nauels A, Xia Y, Bex V, Midgley PM (eds) Contribution of working group I to the fourth assessment report of the intergovernmental panel on climate change. Cambridge University Press, Cambridge, UK/New York

Jamero L, Esteban M, Onuki M (2016) Potential In-Situ adaptation strategies for climate-related sea-level rise: insights from a small island in The Philippines experiencing Earthquake-Induced Land subsidence. J SustaiN 4(2):44–53

Jamero L, Onuki M, Esteban M, Billones-Sensano XK, Tan N, Nellas A, Takagi H, Thao ND, Valenzuela VP (2017) Small island communities in the Philippines prefer local measures to relocation in response to sea-level rise. Nat Clim Chang 7:581–586

Jayaratne MPR, Premaratne B, Adewale A, Mikami T, Matsuba S, Shibayama T, Esteban M, Nistor I (2016) Failure mechanisms and local scour at coastal structures induced by Tsunami. Coast Eng J 58(04)

Keener VW, Marra JJ, Finucane ML, Spooner D (2012) In: Smith MH (ed) Climate change and pacific islands: indicators and impacts. Report for The 2012 Pacific Islands regional climate assessment. Island Press, Washington, DC

Kench PS, McLean RF, Nichol SL (2005) New model of reef-island evolution: Maldives, Indian Ocean. Geology 33(2):145–148

Kench PS, Perry CT, Spencer T (2009) Coral reefs. In: Slaymaker O, Spencer T, Embleton-Hamann C (eds) Geomorphology and global environmental change. Cambridge University Press, Cambridge, pp 180–213

Kouwan Kuukou Gijutsu Kenkyuusho (Port and Airport Research Institute (PARI) 港湾空港技研究所) (2011) Verification of breakwater effects in Kamaishi Ports. http://www.pari.go.jp/info/tohoku-eq/20110401.html. Accessed 19 July 2011 (in Japanese)

McCubbin S, Smit B, Pearce T (2015) Where does climate fit? Vulnerability to climate change in the context of multiple stressors in Funafuti, Tuvalu. Global Environ Chang 30:43–55. http://www.climate.gov.ki/category/action/adaptation/kiribati-adaptation-program/

Mikami T, Shibayama T, Esteban M, Matsumaru R (2012) Field survey of the 2011 Tohoku Earthquake and Tsunami in Miyagi and Fukushima Prefectures. Coastal Eng J 54(1):1–26

Mino T, Esteban M, Anand V, Satanarachchi N, Akiyama T, Ikeda I, Chen C (2016) Philosophy of field methods in the GPSS-GLI program: dealing with complexity to achieve resilience and sustainable societies. In: Esteban M, Akiyama T, Chiasin C, Ikeda I (eds) Sustainability science: field methods and exercises. Springer Nature (forthcoming)

Minoura K, Imamura F, Sugawara D, Kono Y, Iwashita T (2001) The 869 Jogan tsunami deposit and recurrence interval of large-scale tsunami on the Pacific coast of northeast Japan. J Nat Disaster Sci 23(2):83–88

Mori N, Takahashi T, The 2011 Tohoku Earthquake Tsunami Joint Survey Group (2012) Nationwide post event survey and analysis of the 2011 Tohoku Earthquake Tsunami. Coast Eng J 54(1):54–81

Nguyen DT, Takagi H, Esteban M (eds) (2013) Coastal disasters and climate change in vietnam: engineering and planning perspectives. Elsevier, London

Nicholls RJ, Cazenave A (2010) Sea-level rise and its impact on coastal zones. Science 328(5985):1517–1520. https://doi.org/10.1126/science.1185782

Nobuoka H, Murakami S (2011) Vulnerability of coastal zones in the 21st century. In: Sumi A, Mimura N, Masui T (eds) Climate change and global sustainability: a holistic approach. UNU Press, Tokyo

Ogasawara T, Matsubayashi Y, Sakai S, Yasuda T (2012) Characteristics on Tsunami Disaster of Northern Iwate Coast of the 2011 Tohoku Earthquake Tsunami. Coast Eng J 54(1):1250003

Pareschi MT, Favalli M, Boschi E (2006) Impact of the Minoan tsunami of Santorini: Simulated scenarios in the eastern Mediterranean. Geophys Res Lett 33(18):L18607

Perch-Nielsen S, Bättig MB, Imboden D (2008) Exploring the link between climate change and migration. Clim Chang 91:375–393

San Carlos R, Onuki M, Esteban M, Shibayama T (2017) Risk awareness and intended Tsunami Evacuation behaviour of international Tourists in Kamakura City, Japan. Int J Disast Risk Re 23:178–192

Sawai Y, Okamura Y, Shishikura M, Matsuura T, Aung TT, Komatsubara J, Fujii Y (2006) Sendai hirano no taisekibutsu ni kiroku sareta rekishi jidai no kyodai tsunami–1611 nen Keichou tsunami to 869 nen jougan tsunami no shinsui-iki (Historical tsunamis recorded in deposits beneath Sendai Plain -inundation areas of the A.D. 1611 Keicho and the A.D. 869 Jogan tsunamis 仙台平野の堆積物に記録された歴史時代の巨大津波–1611年慶長津波と869年貞観津波の浸水域). Chishitsu Nyuusu (Chishitsu News) 624:36–41. (in Japanese)

Schewe J, Levernmann A, Meinshausen (2011) Climate change under a scenario near 1.5°C of global warming: monsoon intensification, ocean warming and steric sea level rise. Earth Sys Dynam 2(1):25–35

Shibayama T, Esteban M, Nistor I, Takagi H, Nguyen DT, Matsumaru R, Mikami T, Aranguiz R, Jayaratne R, Ohira K (2013) Classification of Tsunami and evacuation areas. J Nat Hazards 67(2):365–386

Simpson C, Kiessling W, Mewis H, Baron-Szabo RC, Müller J (2011) Evolutionary diversification of reef corals: a Comparison of the molecular and fossil records. Evolution 65(11):3274–3284

Solomon S, Qin D, Manning M, Chen Z, Marquis M, Averyt KB, Tignor M, Milller HL (eds) (2007) Climate change 2007: the physical science basis. Contribution of working group I to the 4th assessment report of the intergovernmental panel on climate change. Cambridge University Press, New York

Takagi H, Tran TV, Thao ND, Esteban M (2014) Ocean Tides and the influence of sea-level rise on floods in urban areas of the Mekong Delta. J Flood Risk Manag 8(4):292–300

Tsimopoulou V, Jonkman SN, Kolen B, Maaskant B, Mori N, Yasuda T (2012) A multi-layer safety perspective on the tsunami disaster in Tohoku, Japan. In: Proceedings of flood risk 2012 conference. Rotterdam; 2012

Tsimopoulou V, Vrijling JK, Kok M, Jonkman SN, Stijnen JW (2013) Economic implications of multi-layer safety projects for flood protection. In: Proceedings of the ESREL conference. Amsterdam; 2013

UNISDR (United Nations International Strategy for Disaster Reduction) (2015) Sendai framework for disaster risk reduction 2015–2030. Retrieved May 20, 2017. from http://www.prevention-web.net/files/43291_sendaiframeworkfordrren.pdf

Vermeer M, Rahmstorf S (2009) PNAS 2009;106:21527-21532

Veron JEN, Hoegh-Guldberg O, Lenton TM, Lough JM, Obura DO, Pearce-Kelly P, Sheppard CRC, Spalding M, Stafford-Smit MG, Rogers AD (2009) The coral reef crisis: the critical importance of <350 ppm CO2. Mar Pollut Bull 58(10):1428–1436

Watanabe H (1985) Nihon higai tsunami souran (Comprehensive bibliography on tsunami of Japan 日本被害津波総覧). University of Tokyo Press, Tokyo. (in Japanese)

Webb AP, Kench PS (2010) The dynamic response of reef islands to sea-level rise: Evidence from multi-decadal analysis of island change in the Central Pacific. Glob Planet Chang 72(3):234–246. https://doi.org/10.1016/j.gloplacha.2010.05.003

Yamamoto L, Esteban M (2014) Atoll island states and international law: Climate change displacement and sovereignty. Springer, Heidelberg

Yamamoto L, Esteban M (2016) Migration as an adaptation strategy for Atoll Island States. J Int Migra 55(2):144–158

Yamano H, Sugihara K, Nomura K (2011) Rapid poleward range expansion of tropical reef corals in response to rising sea surface temperatures. Geophys Res Lett 38(4):L04601

Yamao S, Esteban M, Yun NY, Mikami T, Shibayama T (2015) Estimation of the current risk to human damage life posed by future tsunamis in Japan. In: Esteban M, Takagi H, Shibayama T (eds) Handbook of coastal disaster mitigation for engineers and planners. Butterworth-Heinemann (Elsevier), Oxford

Chapter 8
Framing Food Security and Poverty Alleviation

Hirotaka Matsuda, Makiko Sekiyama, Kazuaki Tsuchiya, Chiahsin Chen, Eri Aoki, Rimbawan Rimbawan, and Tai Tue Nguyen

Abstract This chapter addresses current problems of food security, which is considered as one of the most important factors in development strategies to alleviate poverty, by examining relevant policies on agriculture and by discussing different ways to frame food security strategies. The commonly applied framing of development strategies–including strategies for food security–has been the enhancement of market function, even though actual approaches have been changing with the

---

H. Matsuda (✉)
Department of Agricultural Innovation for Sustainability, Faculty of Agriculture,
Tokyo University of Agriculture, Tokyo, Japan
e-mail: hm206784@nodai.ac.jp

M. Sekiyama
Environmental Epidemiology Section, Center for Health and Environmental Risk Research,
National Institute for Environmental Studies, Tsukuba, Ibaraki, Japan
e-mail: sekiyama.makiko@nies.go.jp

K. Tsuchiya
Landscape Ecology and Planning Laboratory, Department of Ecosystem Studies,
The University of Tokyo, Bunkyo, Tokyo, Japan
e-mail: aktcy@mail.ecc.u-tokyo.ac.jp

C. Chen
National Cheng Kung University Museum, National Cheng Kung University,
Tainan City, Taiwan
e-mail: chiahsin@mail.ncku.edu.tw

E. Aoki
Faculty of Information Networking for Innovation and Design, Department of Information
Networking for Innovation and Design, Toyo University, Bunkyo, Tokyo, Japan
e-mail: eriaoki@sfc.keio.ac.jp

R. Rimbawan
Graduate Nutrition Study Program, Dept of Community Nutrition,
Faculty of Human Ecology, Institut Pertanian Bogor/IPB University, Bogor, Indonesia

T. T. Nguyen
Faculty of Geology, VNU University of Science, VNU Key Laboratory of Geoenvironment
and Climate Change Response , Hanoi, Vietnam
e-mail: tuenguyentai@vnu.edu.vn

© The Author(s) 2020   153
T. Mino, S. Kudo (eds.), *Framing in Sustainability Science*,
Science for Sustainable Societies, https://doi.org/10.1007/978-981-13-9061-6_8

recognition of unexpected results such as environmental degradation and nutrition problems. Implementation of these policies has caused a decoupling of production and consumption. As a result, agricultural policies that represent the production or supply side have been implemented apart from nutrition policies, which are relevant to the consumption or demand side. In other words, because agricultural policies and nutrition policies have their own framings and because these framings are not integrated, many problems have occurred. It must be considered how decoupled production or supply can be combined with consumption or demand. In this connection, people's understanding of value that should reflect the shadow price must be transformed. Psychological strategies of providing proper information must be implemented with the SDGs, which are assumed to combine selected aims and targets based on particular contexts by stakeholders such as national governments, local municipalities, private companies, and international organizations.

Keywords Poverty · Food security · Market function · Decoupled production and consumption · Development strategies

8.1 Introduction

This chapter addresses current problems of food security, which is considered as one of the most important factors in development strategies to alleviate poverty, by examining relevant policies on agriculture and by discussing different ways to frame food security strategies. The purpose of development strategies for food security has been primarily to enhance market function. Although, as discussed later, new development strategies have been proposed and implemented after recognizing various newly arising issues to be considered, the above purpose has been essentially unchanged. In other words, the basic framing has remained the same. Development strategies should be supported by corresponding economic theories, which should have been also evolved so that they may support enhancing the market function. The transition of development strategies for food security and poverty alleviation are discussed hereafter along with the supporting economic theories.

Poverty alleviation has been put on the main development agenda at both the national and international levels. As indicated in a series of studies by Amartya Kumar Sen (Sen 1981; Drèze and Sen 1989; Sen 1999), the poverty issue must be investigated from the viewpoints of various dimensions including the amount of goods, services, and income as well as the freedom of utilizing them, people's behavior, and the state of human life. Thus, the framing for poverty alleviation must reflect these various dimensions. The most famous definition of sustainable development was provided in *Our Common Future* by the Brundtland Commission (UN 1987) as follows: *development that meets the needs of the present without compromising the ability of future generations to meet their own needs*. This definition strongly implies the importance of the ethical dimension. The Sustainable Development Goals (SDGs) were announced on January 1, 2016, with 17 aims and

169 targets based on the achievements of the Millennium Development Goals (MDGs) (UN Sustainable Development website and UN 2015). SDGs are the latest and the most comprehensive summary of development strategies. Now, because the SDGs are available as a common action guideline, everyone is encouraged to design strategies in line with the concepts of the SDGs. Development strategies must be designed not only to enhance market functions, but also enhance many other dimensions as described above.

In any overview of development strategies prior to the agreement upon the MDGs in 2000, *structuralism* or structuralist economics (from the 1940s to 1960s) must have been mentioned first (e.g., Singer 1950; Prebisch 1959; Nurkse 1952, 1953). Its fundamental concept was that market functions, including the price mechanism, had not worked in developing countries. Therefore, they insisted that governments had to play an active role to make a "Big Push" for economic development (Rosenstein-Rodan 1943, 1961, 1984). A subsequent approach that relied on neo-classical economics emerged as the mainstream theory of development strategies. The core concept of the *neoclassical approach* was that market functions had worked in both developed countries and developing countries, and that intervention in markets must be strictly limited to occurrences of market failure (e.g., Schultz 1961; Lal 1983; Balassa 1989).

In the 1980s, development strategies aimed to enhance market functions based on the neo-classical economics. Structural adjustment loans (SALs) by the International Monetary Fund and the World Bank (e.g., Hellenier 1987; Williamson 1983; World Bank 1990) are one of the most representative strategies. These development strategies were mainly composed of macro-economic policies. They were encountered by the Basic Needs approach of reformism which centered around the International Labour Organization (ILO), and also by the concept of "adjustment with a human face" by the United Nations International Children's Emergency Fund (UNICEF), which insisted on the importance of viewing poverty at the household or individual level (Oman and Wignaraja 1991, ILO 1976, Hunt 1989; Cornia et al. 1987). In addition, at the beginning of the 1990s, different types of market failures were recognized: (1) poverty; (2) environmental degradation; (3) unequal distribution of resources such as income, property, and food; (4) problems associated with women and children; and (5) infectious diseases such as HIV.

Both formal and informal institutions have played essential roles in responding to these issues. However, development strategies fundamentally remained with the market function approach via neo-classical economic theory, or to some extent a new institutional economy in line with neo-classical economic theory. The core concept of the theory is the way to achieve Pareto Optimum, which is the criterion for the most efficient resource allocation. Because of the two fundamental theorems of welfare economics, Pareto Optimum can be attained under competitive equilibrium and all possible Pareto Optima can be equal to competitive equilibrium through proper wealth redistribution (Arrow 1951; Debreu 1954, 1959; Mas-Colell et al. 1995). In other words, the market mechanism is theoretically guaranteed to play a critical role in achieving efficient resource use and equal allocation of the resource. Achieving perfect market is the pre-condition of the fundamental theorems of welfare economics. Therefore, neo-classical economics was considered to provide the proper rationale to the thought that enhancing market economy can contribute to poverty alleviation, even though the multiple dimensions of poverty were considered important. The approach from the neo-classical economics can be recognized as a framing of development strategy.

In addition to the economic bases as described above, food security has been a central issue in development strategy for poverty alleviation. Many reasons exist for this: (1) the threat of Malthus's world is becoming more of a reality, (2) many people are suffering from poverty in the rural areas of developing countries, (3) agriculture is the main industry in developing countries, (4) the number of undernourished people had remained virtually the same for several decades, and so on. Although the reasons for focusing on food security have changed over time, their fundamental policy has been the same: increasing agricultural productivity in a consistent manner. This policy has remained consistent, even though development strategies have evolved because of recognition of various negative factors such as environmental impacts (IAASTD 2008). Agricultural policies to alleviate poverty from the perspective of food security have been coordinated with economic policies in development strategies. This means that agricultural policy is oriented toward enhancing market function (World Bank 2008).

Enhancing market function decouples production and consumption. When the market function is enhanced, trading is expanded from the household level to local, national, or international levels. People have more potential opportunities to increase their production. Therefore, agricultural policies have primarily focused on the production or supply side. On the other hand, consumption is more related to nutrition policy. This decoupling of consumption from production often leads to not only the market failure but also mal-nutrition, and thus causes failure in poverty alleviation policies by the government. Agricultural policy dealing with production and nutrition policy dealing with consumption have different framings. Both policies are implemented separately and not integrated with each other because of decoupling consumption from production that results from development strategies.

In the following sections, we first give an overview of the impacts of the Green Revolution on poverty alleviation in developing countries. The Green Revolution began in the 1940s, and the development strategies associated with the Green Revolution were to enhance market function in agriculture. It impacted food security in the world tremendously through the decoupling of consumption from production. Second, we review how nutrition status was considered in development strategies. The impacts of the Green Revolution are discussed from the viewpoint of the production or supply side perspective. Nutrition status, however, is discussed from the viewpoint of consumption or demand side perspective. Third, we discuss the reasons for the failure of the market economy concept in neo-classical economic theory, as well as its implications. A summary of this chapter is provided at the end.

8.2 Impacts of the Green Revolution on Developing Countries

8.2.1 Brief History of the Green Revolution

The Green Revolution was named by William Gaud (1968), a former administrator of the United States Agency for International Development (USAID). It was a process of increasing the yields of cereals, wheat, and rice by developing a high yield variety (HYV) or modern variety (MV) with modern inputs such as modern irrigation systems, chemical fertilizers, and pesticides. It appeared in 1940s, and proceeded from the 1960s to the late 1980s or early 1990s. Developments of HYV of wheat and rice marked the beginning of the Green Revolution. In the early 1940s, the whole world stood in awe of reality of the threat of Malthus's world. In 1798, Thomas Robert Malthus published a famous book, *An Essay on the Principle of Population,* and insisted in the book that human population increased in a geometric or exponential manner, whereas the ability to produce food increased only in an arithmetic manner. In other words, the population would outgrow the ability of the land to produce food. Food shortages would have been real.

Agronomist Dr. Norman Earnest Borlaug, the recipient of the Nobel Peace Prize in 1970 and often known as the "Father of the Green Revolution," succeeded in developing HYV wheat at an institution in Mexico in 1943 under a variety improvement operation by the Rockefeller foundation, which had taken the initiative to counter potential food shortages. In the Mexican institute, the HYV wheat was developed by crossing Norin 10 from Japan onto an indigenous variety of spring wheat in Mexico. The developed HYV wheat variety was defused to Asian countries. To bring varieties from other areas, adaptation to differences in natural conditions was required. In the process of its diffusion among Asian countries, it was crossed onto area varieties through collaboration with National Agricultural Research Systems (NARS).

Agricultural research and development (R&D) systems, including NARS, were established through the process. The institute in Mexico became the International Maize and Wheat Improvement Center, CIMMYT (Centro Internacional de Mejoramiento de Maíz y Trigo), which is an institution under CGIAR (Consultative Group on International Agricultural Research). In the case of the other main cereal *rice*, HYV rice was developed in an institute in the Philippines, which became an institute under CGIAR, IRRI (International Rice Research Institute) supported by the Ford Foundation after the Rockefeller Foundation. One of the representative types of HYV rice was IR8, which was developed in 1966. It was produced by crossing semi-dwarf rice varieties from Taiwan onto a rice variety from Indonesia in order to respond to chemical fertilizer and increase productivity.

Figure 8.1 indicates yield trends by region and compares yield increase between Asia and Africa. In the 1960s, yields of cereal were virtually the same in South Asia and Sub-Saharan Africa. However, the difference has expanded since the 1980s, which came after the Green Revolution. In addition, when comparing the relationship

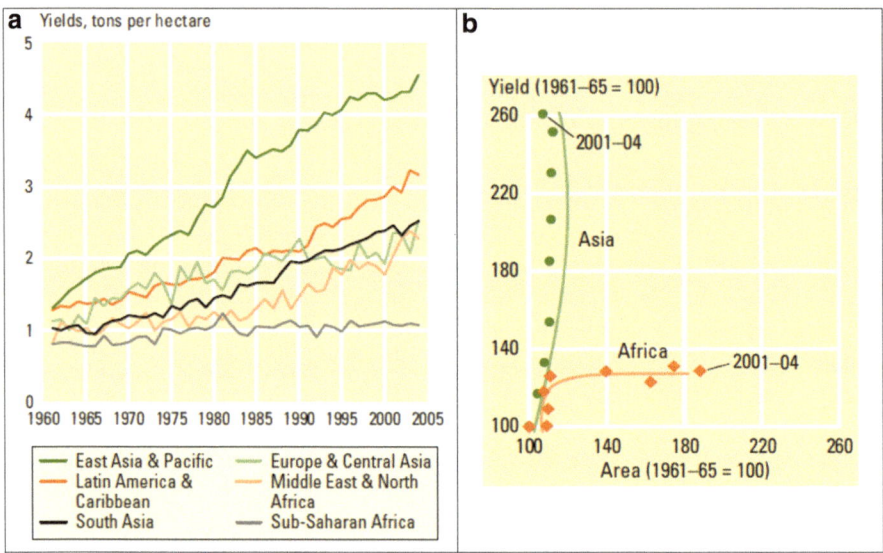

Fig. 8.1 Regional yield growth. (**a**) Trend of yield growth (Source: World Bank (2008)). (**b**) Yield growth in Asia and Africa (Source: FAO 2006)

between yield increase and area expansion of farm land, the increase in yield in sub-Saharan Africa has been stagnant in spite of the expansion of farm land while Asia increased cereal yield sharply without expanding farm land.

When Fig. 8.2 is examined again, it can be found that HYV and inputs other than farm area significantly contributed to the increase in cereal yield in Asia. Agricultural lands, including both croplands and pastures, occupy approximately 38% of Earth's terrestrial surface; it is the largest land use on the planet (Foley et al. 2011). The expansion of agricultural land is often preceded by deforestation (Lambin and Meyfroidt 2011), which has greatly influenced various environmental issues such as increase in greenhouse gas emissions, destruction of biological habitats, and decline in ecosystem services (Bommarco et al. 2013). In addition, because of restrictions on the expansion of agricultural land use, it has been intensified through the introduction of chemical fertilizers and pesticides, and through an increase in irrigation capacity to increase yield.

Agricultural land intensification impacts climate to a similar extent as land use changes (Luyssaert et al. 2014) and results in changes to local, regional and global biogeochemical and water cycles. It likewise has had major influence on biodiversity loss (Erb et al. 2013). Because choice of agricultural land management method greatly impacts the environment, interest in sustainable intensification has been increasing (Garnett et al. 2013; Tilman et al. 2011). Other inputs can be considered so-called modern inputs such as modern irrigation systems, chemical fertilizers, and pesticides. Though there are many reasons for the different results of various industrial policies, the differences observed between countries in Asia and in Sub-Saharan

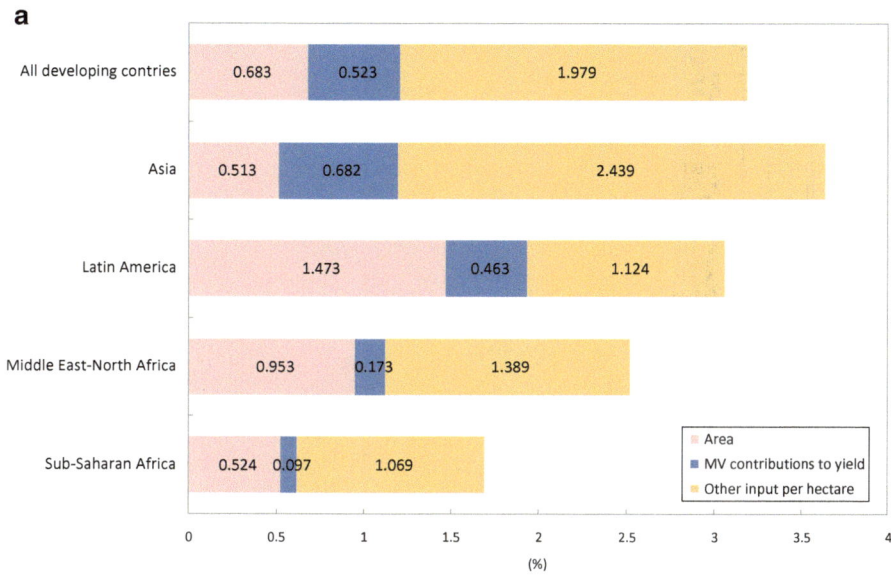

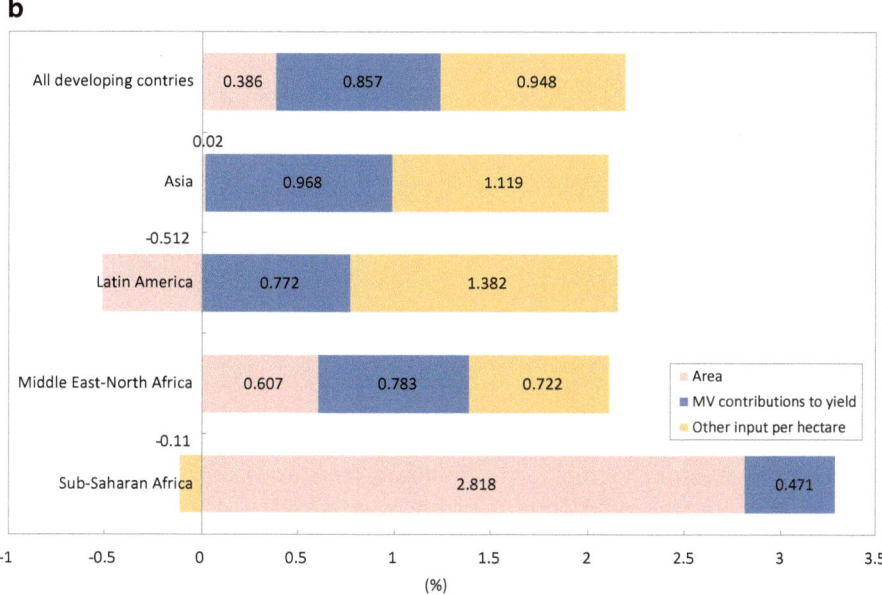

Fig. 8.2 Decompose of factors for yield growth. (**a**) Early Green Revolution Period, 1961–1980. (**b**) Late Green Revolution Period, 1981–2000. (Source: Made by author from Evenson and Gollin (2003))

Africa are not discussed in detail because such discussion goes beyond the main purpose of this chapter. However, let it suffice to say that the chemical industry could be fostered in Asia to provide rather cheap chemical fertilizer. The Green Revolution can be considered as a process to make sufficient preparation to enhance market mechanism in agriculture. Details of the process have been discussed from the perspective of the meaning of the Green Revolution in the context of development strategy in the next section.

8.2.2   The Meaning of the Green Revolution in the Context of Development Strategy

The Green Revolution introduced the opportunity of applying the accumulated knowledge of science to agriculture. The Green Revolution preceded the development of HYV by purposefully breeding to enhance its yield with chemical fertilizers. Narrowing the gap between production in the actual field and the experiment field was made possible by the development of HYV, and by the accumulation of scientific knowledge such as the utilization of modern irrigation systems and the appropriate use of chemical fertilizers and pesticides. In another perspective, private sector actors (such as the chemical industry and the medical industry), who can use their accumulated knowledge and skill to produce pesticides, were able to take part in the agricultural industry after the Green Revolution. Market economics intruded into the agricultural industry after the Green Revolution. International organizations such as FAO supported this change by expanding international seed operations, i.e., International Seed Campaign (1957–1962), World Seed Year (1961), Improved Seed Development Plan (1973), among others.

Importantly, multinational companies entered the market. Multinational chemical companies and medical companies entered the seed industry. Multinational agribusiness companies together with food distribution and food processing companies also entered the seed industry. In addition, advances in breeding technology caused increased R&D investment and an increasing gap between large and small seed companies. Then, aggravation of M&A in the seed industry was caused by multinational companies who concentrated on a very small number of companies to address seed development and production in order to lower their risk. The multinational companies bought up existing seed companies rather than starting their own seed operations as newcomers, which would have required them to obtain genetic resources and acquire their own breeding technology to compete with other competitors in the market. In any case, the market economy penetrated and expanded into agriculture after the Green Revolution as a development strategy to alleviate poverty.

However, the Green Revolution also caused several problems of market or government failure, as did other aforementioned development strategies. High intensive input farming, which is one of features of the Green Revolution, typically involves monocrop fields and a package of modern seed varieties, fertilizers, and pesticides. It causes various problems such as water pollution, indirect damage to larger eco-systems, and inadvertent pesticide poisoning of humans, animals, and non-targeted plants and insects caused by mismanagement of irrigation water, injudicious use of fertilizers and pesticides, and excess chemical fertilizer use (World Bank 2008). Figure 8.3 indicates the growth of concern in agriculture.

With the various problems caused by the Green Revolution, as with other devel-opment strategies to alleviate poverty, unexpected and new issues have arisen and been recognized over time. In other words, these issues were from that framing in the beginning had been not enough. One of the important features of the market function under neo-classical economic theory detaches demand or consumption from supply or production. This can also be found in agriculture after the Green Revolution. Although problems caused after the Green Revolution are commonly recognized, those are only perspectives from supply or production.. The framing of development strategies in food security from the production perspective has changed to recognize unexpected impacts such as environmental degradation, although the framing of poverty alleviation, which is a general development strategy to enhance market function in food security, remained. To understand the features and results of development strategy framing, particularly enhancing food security to alleviate poverty, those perspectives are still insufficient. Although if all the market failures in production or supply side have been confronted properly, such efforts would still be insufficient to alleviate poverty. Another framing is needed for poverty allevia-tion in terms of food security. Demand side must be considered. The impact of enhancing market function in demand or consumption of agricultural production

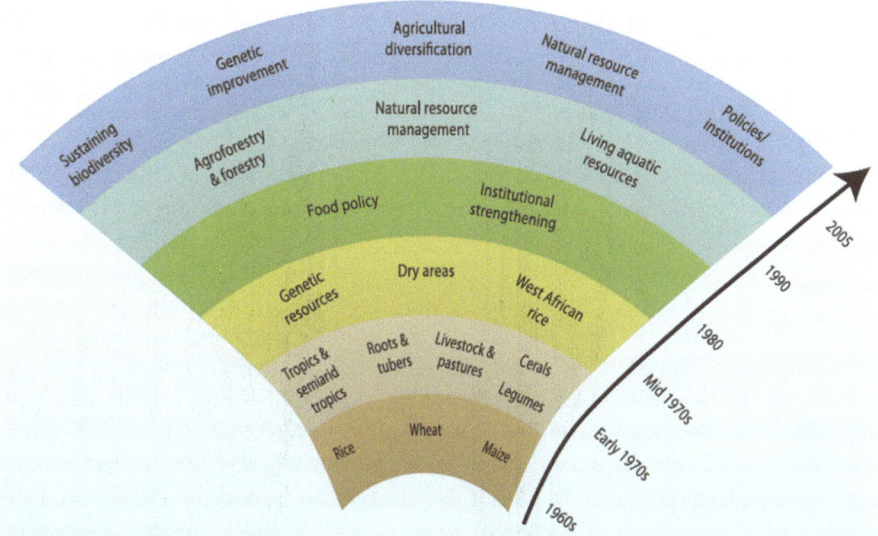

Fig. 8.3 Pathway to the current conception of modern agriculture. (Source: IAASTD (2008))

after the Green Revolution—in short, nutrition status—is discussed in next section.

8.3 Poverty Alleviation from a Nutrition Perspective

8.3.1 Development Strategies for Nutrition

Framing food security from the perspective of demand has changed with the recognition that previously implemented strategies failed and led to unexpected results, as is the case with the production perspective. In this section, transition of development strategies from a nutrition perspective is overviewed first.

The MDGs made poverty and hunger alleviation as their first target, aiming "to halve between 1990 and 2015 the proportion of people who suffer from hunger" (UN Millennium Project 2005). Global food production was more than enough to feed the global population, however 14% of the world population, including children, was undernourished in 2007 (FAO 2008). Nearly nine million children under the age of five died in 2007, and over one-third of those deaths were linked to undernutrition (FAO 2008). Consequently, it became recognized that alleviation of child undernutrition was the most unattainable goal among the MDGs because of the insufficient effort of the countries involved and the international community.

Under such a situation, the Lancet series on maternal and child undernutrition was launched in 2008, pointing out the urgent need to scale up the international nutrition governance system (Horton et al. 2008). The series also highlighted the short-term and long-term consequences of child undernutrition: child undernutrition affects not only disability, morbidity, and mortality during his/her childhood stage but also affects body size, intellectual ability, economic productivity, reproductive performance, and cardiovascular disease risk during his/her adulthood stage (Black et al. 2008). The final paper of this series stated that 'the international nutrition system comprised of international and donor organizations, academia, civil society, and the private sector–is fragmented and dysfunctional' (Morris et al. 2008). Provoked by this series, political commitment to child malnutrition increased (Table 8.1).

In 2010, Scaling Up Nutrition (SUN) was established to tackle stunting (low height-for-age, indicating chronic restriction of a child's potential growth) with a special focus on the 1000-day window of opportunity, from conception to a child's second birthday. The SUN movement currently has 54 member countries (Menon et al. 2014). In 2012, WHO issued a resolution at the 65th World Health Assembly endorsing a comprehensive implementation plan on maternal, infant, and young child nutrition that specified a set of six global nutrition targets by 2025. Importantly, one of those six targets was 'to ensure that there is no increase in childhood overweight', which was the first attempt to regard 'overnutrition' as an increasing form of malnutrition (WHO 2012). In 2013, first Nutrition for Growth (N4G) summit was

Table 8.1 Global movement on nutrition issues

Jan 2008	1st series of maternal and child undernutrition in Lancet	Highlight the importance of first 1000 days for tackling malnutrition.
Sep 2010	Scaling Up Nutrition (SUN) initiated	
May 2012	65th World Health Assembly was held. 'comprehensive implementation plan on maternal, infant and young child nutrition' was adopted.	Global Nutrition Targets 2025 was decided.
June 2013	First Nutrition for Growth (N4G) summit was held.	
Aug 2013	2nd series of maternal and child undernutrition in lancet	
Nov 2014	FAO and WHO jointly organized Second International Conference on Nutrition	Focused on targeting all forms of malnutrition.
Sep 2015	SDGs	12 out of 17 targets included nutrition related issues

hosted by the governments of Brazil and the UK, yielding over 200 commitments to expand the reach of nutrition interventions during the 1000-day window, reducing stunting, and saving the lives of almost two million children under the age of five. Two months after the first N4G, the second Lancet series on maternal and child nutrition was published; in this series, the title was changed from 'undernutrition' to 'overnutrition', highlighting the increasing awareness of dual forms of malnutrition (Black et al. 2013).

In 2014, FAO and WHO jointly organized the second International Conference on Nutrition, which had a strong focus on tackling all forms of malnutrition including undernutrition, micronutrient deficiencies, and overnutrition with a view to achieving global nutrition targets set by the World Health Assembly by 2025 (Haddad et al. 2015). To this end, unlike the MDGs framework, the 2013 High Level Panel on the Post 2015 Development Agenda explicitly recommends nutrition as an explicit feature of one of its proposed goals (Haddad et al. 2015). Consequently, in the Sustainable Development Goals (SDGs) launched in 2015, 12 out of 17 targets included nutrition-related issues (IFPRI 2016).

UNICEF framed the nutritional status of children as determined by a range of immediate, underlying, and basic causes (UNICEF 1990), and different approaches addressing different levels of those determinants have been taken to tackle child malnutrition problem. Nutrition-specific interventions address the immediate determinants (i.e., dietary intake and disease) through micronutrient supplementation and fortification, improvement of breastfeeding and complementary feeding, and strengthening hygiene. Most of the intervention programs have been in the form of a nutrition-specific approach, which is highly project- or donor-dependent and is difficult for the local community to sustain.

A nutrition-sensitive approach addresses underlying determinants (i.e., food security, health access, healthy household environment, and care practices) through improving agriculture, water and sanitation, education, and social protection (Black et al. 2013). Global interest in this nutrition-sensitive approach, which might offer sustainable solutions for local population, is recently increasing. However, so far, the evidence base is weak on how to make interventions that address more nutrition-sensitive underlying determinants (Haddad et al. 2015). Masset et al., for example, systematically reviewed papers assessing the impact of agricultural intervention on improvements in child nutritional status and concluded that very little evidence was found for such intervention, because the prevalence of malnutrition was mostly caused by methodological weakness (2012).

The Global Nutrition Reports 2014 and 2015 provided some specific ideas for interventions on food systems and agriculture; however, evidence is limited to kitchen gardening and bio-fortification (IFPRI 2016). Furthermore, as mentioned earlier, global efforts to combat child malnutrition have until recently mainly targeted undernutrition. Many developing countries now, however, need to simultaneously solve dual forms of malnutrition such as undernutrition and overnutrition. Child overweight or obesity is rapidly increasing in those countries (Black et al. 2013). The basic driver of undernutrition is mostly limited food accessibility, availability, and affordability, whereas that of overnutrition is basically inappropriate dietary choices or behaviors.

Inappropriate dietary choices or behaviors relate to various socioeconomic factors such as subsistence patterns, local economy, food system, food environment, and health and nutrition education, which imply that multi-sectoral and transdisciplinary approaches are necessary to solve the problem. Solving the problem of undernutrition used to be the main target of development strategies in the context of food security. In terms of quantity of food, agriculture enhanced by development strategies after the Green Revolution contributed to food security. However, there have been many problems in terms of quality and nutrition intake. In addition, the unexpected issue of becoming overweight is now arising. At the same time, it has been recognized that dietary patterns impact environment, which is discussed in the next section.

8.3.2 Environment and Nutrition

In 2010, FAO proposed the concept of a 'sustainable diet' – a diet with low environmental impact that contributes to food and nutrition security and to a healthier life for present and future generations (FAO 2012). Since that time, the amount of literature assessing the environmental impact of dietary behavior has been increasing. The review work of Auestad and Fulgoni about 'sustainable diet' found 31 papers examining sustainable diet mostly published from 2010 to 2014. Those papers assessed different indicators of the environmental impacts of different diet patterns such as greenhouse gas (GHG) emissions, land capacity, energy/fossil fuel use, and water use (Auestad and Fulgoni 2015).

The health community also recognizes that environmental sustainability is increasing. WHO recently highlighted the importance of win-win approaches to health and environment, especially those that reduce GHG emissions and increase resilience to environmental change. For example, reducing saturated fat intake from livestock products has co-benefits for health and environment: 8–9% of global GHG emissions come from the livestock sector, which provides large amounts of saturated fat leading to increased risk of cardiovascular diseases (Friel et al. 2009). WHO recommends high saturated-fat, high calorie-meats and processed foods be substituted with more unprocessed foods, fiber-rich foods, and fresh fruits and vegetables. The Global Nutrition Report also recommends increasing dietary diversity, especially increasing fruits and vegetables in the diet (IFPRI 2016). Many developing countries, however, are shifting from subsistence farming to industrialized food production, leading to the gradual loss of biodiversity in local food varieties of legumes, fruits, nuts, seeds, and berries, and to an increased reliance upon simplified diets of imported food varieties or mass-produced staples. This change has led to diets that are energy rich but contain few vital micronutrients. Thus, more effort is urgently needed to provide locally sustainable solutions to improve nutrition and agricultural systems.

In recent years, data show how changes in dietary habits strongly affect health and the environment (Aleksandrowicz et al. 2016; Springmann et al. 2016; Tilman and Clark 2014). Erb et al. (2016) have been studying the effects of dietary habits at a global level and have shown that it is possible to nurture a growing human population without deforestation by combining the transition to a vegetarian diet and sustainable intensification. These scenario-based studies provide insights into potential pathways for food systems change. These studies, however, do not tell much about what policies or interventions will help to realize the illustrated scenarios. Because most of the future population growth is expected to occur in cities, it will be necessary to prepare a nutritious and environmentally-friendly food environment in newly developed areas and to transform the food environments of existing urban areas.

Enhancing market function was the main development strategy to alleviate poverty. As a consequence, decoupling of consumption from production occurred, followed by unexpected market failure and government failure as explained in detail in the next section. These undesirable situations were caused by an improper framing, where production/supply and consumption/demand were separately handled in the policy application. Namely, the food supply (production) was governed by neoclassical economic theories without consideration of environmental factors; and the nutritious demand for food (consumption) did not properly consider people's diet patterns. In fact, people's psychological response for the applied strategies/policies, preference for food, perception and values about food, among others are essential factors when actual strategies/policies are designed and implemented. In the next section, theoretical framework of people's value in the supporting theory of development strategies is overviewed.

8.4 Dysfunction of Framing in Poverty Alleviation

8.4.1 Market Failure and Shadow Prices

The market failure was a consequence of dysfunction in the market economy framing for poverty alleviation, and caused by several factors including externalities, public goods, scale economy, asymmetric information, and uncertainties. As explained in the previous section, the Green Revolution unearthed various unexpected environmental degradation issues. The dissociation of actual prices and shadow prices is one reason for this. *Shadow price* refers to the true social value or cost of a resource. Neo-classical economic theory assumes that people decide their behavior based on a sense of value, which is reflected on the price in the market through subjectivity to maximize utility.

Prices must reflect the real value of goods including the impacts on the environment. If a price does not reflect the real value, the resource may not be used properly and environmental degradation will occur. People should be aware of the mechanism of shadow price so that they can behave and use resources properly. The difficulty of estimating the shadow price of natural resources is often highlighted, but one reason for unexpected environmental degradation is that people behave without understanding the proper shadow price of the resource. Furthermore, shadow price can change dynamically by transforming the resources' sense of value in the society. In the implementation process of development strategies to alleviate poverty, food consumption or nutrition intake was detached from agricultural production after the Green Revolution. The pursuit of increasing agricultural productivity led to monocrop agriculture, which means transition from subsistence farming to industrialized food production. Under such a situation, malnutrition still coexists with overnutrition at household, local, regional, national and global levels. Overnutrition results from improper dietary choice or behavior based on the value reflected by shadow prices. The value of a resource in a society is often formulated by insufficient knowledge or information, and people cannot behave properly with improper knowledge. In the next section, potential interventions are discussed.

8.4.2 Social and Behavior Change

Behavioral theories and models aim to extract psychological structures common to many behaviors, and to describe them as usually focusing on some specific core factors. However, human behaviors are often decided through many steps and various influential factors: intention, attitude, locust of control, self-efficacy, knowledge, opportunity, cultural habitats, social supports, among others. In addition, social determinants including knowledge, attitudes, social norms, and cultural practices from the individual level to the society level should affect the behaviors. Social and behavioral change communication (SBCC) programs have been well known as powerful tools for fundamental understanding of human interaction (communication) that can strongly influence the social dimensions of health and well-being. According to SBCC, evidence-based communication programs can increase knowledge, shift attitudes and cultural norms, and produce changes in a wide variety of behaviors (Lamstein et al. 2014). Obtaining and providing scientific knowledge about these factors based on research is the first step of great importance, although some uncertainties about human behaviors always exist. Furthermore, SBCC emphasizes that communication goes beyond the delivery of a message, and encompasses the full range of ways in which people can individually–as well as collectively–identify the meanings of their behaviors, and raise the level of impacts on social and behavioral changes.

In a way, this type of change has a wider coverage than individual change. SBCC is the systematic application of interactive, theory-based, and research-driven communication processes that provides strategies for change at three levels: individual, community, and society. There are three main strategies of SBCC across these levels: (1) behavior change communication that applies multimedia and participatory approaches for change at the individual and community levels; (2) social and community mobilization that develop partnerships and alliances to influence the community level to the national level, and (3) advocacy that bridges different levels of approaches and their influences through political and social commitment. It is important to involve multiple stakeholders in a knowledge exchange and create platforms for that in order to design, implement and disseminate the scientific evidence. Multiple stakeholders broadly encompass local governments, local NPO/NGO, private sector, local residents, scientists, researchers, and mass media, including others who have been working on local issues.

Fujii et al. (2001) and Fujii (2016) proposed concepts of structural and psychological strategy for planning transportation systems, which implied many political issues. Structural strategy can be considered as a method for changing the social structure, which corresponds to enhancing market structure and reducing market and government failure. Psychological strategies influence psychological factors such as morale or recognition without changing the social structure. Providing proper information is part of this strategy. People's values define their behavior, and are reflected in shadow prices. However, formulated values are not always proper from various perspectives in development strategies and the value in itself must change as society changes. Thus, psychological strategy plays an important role alongside structural strategy, and the psychological approach can be applied to development strategies, including strategies relevant to food security.

8.5 Summary

Food security was, and still is, a main issue in development strategies to alleviate poverty. The commonly applied framing of development strategies–including strategies for food security–has been the enhancement of market function, even though actual approaches have been changing with the recognition of unexpected results such as environmental degradation and nutrition problems. Agricultural policies that represent the production or supply side have been implemented apart from nutrition policies, which are relevant to the consumption or demand side. In other words, implementation of these policies has caused a decoupling of production and consumption. Because agricultural policies and nutrition policies have their own framings and because these framings are not integrated, many problems have occurred. Livelihoods used to rely on natural resources in the past, whereas consumption of food or nutrition intake is now detached from production or supply of food at household, regional, national, and international levels. People who used to live in self-sufficient economies now begin to purchase necessary goods and services with money they have earned by using their own resources including their own labor at the household level. At the same time, this transformation of the economy at the household level is enhanced through cross-regional and international trading.

Because of the multidimensional perspectives of poverty, enhancing market function as a general principle of development strategies–including strategies relevant to food security–appears a rational and palatable approach. Thus, structural strategies have been applied to reduce market failure and government failure. It must be considered how decoupled production or supply can be combined with consumption or demand. In this connection, people's understanding of value that should reflect the shadow price must be transformed. Psychological strategies of providing proper information must be implemented with the SDGs, which are assumed to combine selected aims and targets based on particular contexts by stakeholders such as national governments, local municipalities, private companies, and international organizations.

References

Aleksandrowicz L, Green R, Joy EJ, Smith P, Haines A (2016) The impacts of dietary change on greenhouse gas emissions, land use, water use, and health: a systematic review. PLoS One 11(11):e0165797

Arrow KJ (1951) An extension of the basic theorems of classical welfare economics. In: Proceedings of the second Berkeley symposium on mathematical statistics and probability. University of California Press, Berkeley/Los Angeles, pp 507–532

Auestad N, Fulgoni VL 3rd (2015) What current literature tells us about sustainable diets: emerging research linking dietary patterns, environmental sustainability, and economics. Adv Nutr 6(1):19–36

Balassa B (1989) Outward orientation. In: Chenery H, Srinivasan TN (eds) Handbook of development economics, vol 2(2). Elsevier, Atlanta, pp 1645–1689

Black RE, Allen LH, Bhutta ZA, Caulfield LE, de Onis M, Ezzati M, Mathers C, Rivera J for the Maternal and Child Undernutrition Study Group (2008) Maternal and child undernutrition: global and regional exposures and health consequences. Lancet 371(9608):243–260

Bommarco R, Kleijn D, Potts SG (2013) Ecological intensification: harnessing ecosystem services for food security. Trends Ecol Evol 28(4):230–238

Cornia J, Jolly R, Stewart F (eds) (1987) Adjustment with a human face: protecting the vulnerable and promoting growth, 2 vols. Clarendon Press, Oxford

Crutzen PJ (2002) Geology of mankind. Nature 415:23. https://doi.org/10.1038/415023a

Debreu G (1954) Valuation equilibrium and pareto optimum. Proc Natl Acad Sci U S A 40(7):588–592

Debreu G (1959) Theory of value. Wiley, New York

Drèze J, Sen A (1989) Hunger and public action. Clarendon Press, Oxford

Erb KH, Haberl H, Jepsen MR, Kuemmerle T, Lindner M, Müller D, Verburg PH, Reenberg A (2013) A conceptual framework for analysing and measuring land-use intensity. Curr Opin Env Sust 5(5):464–470

Erb KH, Lauk C, Kastner T, Mayer A, Theurl MC, Haberl H (2016) Exploring the biophysical option space for feeding the world without deforestation. Nat Commun 7(11382):1–7. https://doi.org/10.1038.ncomms11382

Evenson RE, Gollin D (2003) Assessing the impact of the Green Revolution, 1960 to 2000. Science 300(5620):758–762

FAO (2006) AOSTAT. Rome

FAO (2008) The state of food insecurity in the world 2008. FAO, Rome

FAO (Food and Agricultural Organization) (2012) Proceedings of the international scientific symposium, biodiversity and sustainable diets 2010. FAO, Rome

Foley JA, Ramankutty N, Brauman KA, Cassidy ES, Gerber JS, Johnston M, Mueller ND, O'Connell C, Ray DK, West PC, Balzer C, Bennett EM, Carpenter SR, Hill J, Monfreda C, Polasky S, Rockström J, Sheehan J, Siebert S, Tilman D, Zaks DPM (2011) Solutions for a cultivated planet. Nature 478:337–342

Friel S, Dangour AD, Garnett T, Lock K, Chalabi Z, Roberts I, Butler A, Butler CD, Waaqe J, McMichael AJ, Haines A (2009) Public health benefits of strategies to reduce greenhouse-gas emissions: food and agriculture. Lancet 374(9706):2016–2025

Fujii S (2016) Prescription for social dilemmas: psychology for urban, transportation, and environmental problems. Springer, Tokyo

Fujii S, Gärling T, Kitamura R (2001) Changes in drivers' perceptions and use of public transport during a freeway closure: effects of temporary structural change on cooperation in a real-life social dilemma. Environ Behav 33(6):796–808

Garnett T, Appleby MC, Balmford A, Bateman IJ, Benton TG, Bloomer P, Burlingame B, Dawkins M, Dolan L, Fraser D, Herrero M, Hoffmann I, Smith P, Thornton PK, Toulmin C, Vermeulen SJ, Godfray HCJ (2013) Sustainable intensification in agriculture: premises and policies. Science 341(6141):33–34

Griggs D, Stafford-Smith M, Gaffney O, Rockström J, Öhman MC, Shyamsundar C, Shyamsundar P, Steffen W, Glaser G, Kanie N, Noble I (2013) An integrated framework for sustainable development goals. Ecology and Society 19(4):49

Haddad L, Achadi E, Bendech MA, Ahuja A, Bhatia K, Bhutta Z, Blössner M, Borqhi E, de Onis M, Eriksen K, Fanzo J, Flores-Ayala R, Fracassi P, Kimani-Murage E, Nago Koukoubou E, Krasevec J, Newby H, Nugent R, Oenema S, Matrin-Prével Y, Randel J, Requejo J, Shyam T, Udomkesmalee E, Reddy KS (2015) The global nutrition report 2014: actions and accountability to accelerate the world's progress on nutrition. J Nutr 145(4):663–671

Hellenier GK (1987) Stabilization, adjustment, and the poor. World Dev 15(12):1499–1513

Horton R (2008) Maternal and child undernutrition: an urgent opportunity. Lancet 371(9608):179

Hunt D (1989) Economic theory of development: an analysis of competing paradigms. Harvester Wheatsheaf, New York

IAASTD (International Assessment of Agricultural Knowledge, Science and Technology for Development) (2009) Agriculture at a crossroads (McIntyre BD, Herren HR, Wakhungu J, Watson RT (eds)). Island Press, Washington, DC

IFPRI (International Food Policy Research Institute) (2016) Global nutrition report 2016: actions and accountability to advance nutrition and sustainable development. IFPRI, Washington, DC. Available at http://www.globalnutritionreport.org

ILO (International Labor Organization) (1976) Employment, growth and basic needs: a one-world problem. ILO, Geneva

Lal D (1983) The poverty of 'development economics'. Institute of Economic Affairs, London

Lambin EF, Meyfroidt P (2011) Global land use change, economic globalization, and the looming land scarcity. Proc Natl Acad Sci U S A 108(9):3465–3472

Lamstein S, Stillman T, Koniz-Booher P, Aakesson A, Collaiezzi B, Williams T, Beall K, Anson M (2014) Evidence of effective approaches to social and behavior change communication for preventing and reducing stunting and anemia: report from a systematic literature review. USAID/ Strengthening Partnerships, Results, and Innovations in Nutrition Globally (SPRING) Project, Arlington

Luyssaert S, Jammet M, Stoy PC, Estel S, Pongratz J, Ceschia E, Churkina G, Don A, Erb K, Ferlicoq M, Gielen B, Grünwald T, Houghton RA, Klumpp K, Knohl A, Kolb T, Kuemmerle T, Laurila T, Lohila A, Loustau D, McGrath MJ, Meyfroidt P, Moors EJ, Naudts K, Novick K, Otto J, Pilegaard K, Pio CA, Rambal S, Rebmann C, Ryder J, Suyker AE, Varlagin A, Wattenbach M, Dolman AJ (2014) Land management and land-cover change have impacts of similar magnitude on surface temperature. Nat Clim Chang 4:389–393

Mas-Colell A, Whinston MD, Green JR (1995) Microeconomic theory. Oxford University Press, Oxford

Masset E, Haddad L, Cornelius A, Isaza-Castro J (2012) Effectiveness of agricultural interventions that aim to improve nutritional status of children: systematic review. BMJ 344:d8222

Morris SS, Cogill B, Uauy R, Maternal and Child Undernutrition Study Group (2008) Effective international action against undernutrition: why has it proven so difficult and what can be done to accelerate progress? Lancet 371(9612):608–621

Nurkse R (1952) Some international aspects of the problem of economic development. Am Econ Re 42(2):571–583

Nurkse R (1953) Problems of capital formation in under-developed countries. Basil Blackwell, Oxford

Oman CP, Wignaraja G (1991) The postwar evolution of development thinking. Palgrave Macmillan, London

Prebisch R (1959) Commercial policy in the underdeveloped countries. Am Econ Rev 49(2):251–273

Rosenstein-Rodan PN (1943) Problems of industrialization of eastern and south-eastern Europe. Econ J 53(210/211):202–211

Rosenstein-Rodan PN (1961) Notes on the theory of the 'Big Push'. In: Ellis HS (ed) Economic development for Latin America. Macmillan, London, pp 57–81

Rosenstein-Rodan PN (1984) Natura facit saltum: analysis of the disequilibrium growth process. In: Meier GM, Seers D (eds) Pioneers in development. Oxford University Press, New York, pp 205–221

Sen A (1981) Poverty and famines: an essay on entitlement and deprivation. Clarendon Press, Oxford

Sen A (1999) Development as freedom. Oxford University Press, New York

Shultz TW (1961) Investment in human capital. Am Econ Rev 51(1):1–17

Singer HW (1950) The distribution of gains between investing and borrowing countries. Am Econ Rev 40(2):473–485

Springmann M, Godfray HCJ, Rayner M, Scarborough P (2016) Analysis and valuation of the health and climate change cobenefits of dietary change. Proc Natl Acad Sci U S A 113(15):4146–4151

Steffen W, Richardson K, Rockström J, Cornell SE, Fetzer I, Bennett EM, Biggs R, Carpenter SR, de Vries W, de Wit CA, Folke C, Gerten D, Heinke J, Mace GM, Persson LM, Ramanathan V, Reyers B, Sörlin S (2015) Planetary boundaries: guiding human development on a changing planet. Science 347(6223)

Tilman D, Clark M (2014) Global diets link environmental sustainability and human health. Nature 515(7528):518–522

Tilman D, Baltzer C, Hill J, Befort BL (2011) Global food demand and the sustainable intensifica-
    tion of agriculture. Proc Natl Acad Sci U S A 108(50):20260–20264
UN Millennium Project (2005) Halving hunger: it can be done. Task Force on Hunger, London/
    Sterling. United Nations Sustainable Development Homepage, http://www.un.org/sustain-
    abledevelopment/. Accessed on Aug 2017
UNICEF (United Nations Children's Fund) (1990) Strategy for improved nutrition of children and
    women in developing countries. Policy Review Paper E/ICEF/1990/1.6, UNICEF. New York;
    JC 27/UNICEF-WHO/89.4. New York
United Nations (2015) The millennium development goals report
United Nations World Commission on Environment and Development (1987) Our common future.
    Report of the World Commission on Environment & Development
WHO (World Health Organization) (2012) Resolution WHA65.6. Maternal, infant and young
    child nutrition. In: Sixty-fifth World Health Assembly, Geneva, 21–26 May. Resolutions and
    secisions, annexes. Geneva, World Health Organization. (WHA65/2012/REC/1). Available at:
    http://apps.who.int/gb/ebwha/pdf_files/WHA65/A65_R6-en.pdf. Accessed 29 Nov 2016
Williamson J (1983) The lending policies of the International Monetary Fund. In: Williamson
    J (ed) IMF conditionality. Institute of International Economics, Washington, DC
World Bank (1990) World development report 1990. Oxford University Press, Oxford
World Bank (2008) World development report 2008. Oxford University Press, Oxford

Part III
Epilogue

Chapter 9
Linking Framing to Actions for Sustanability

Takashi Mino and Shogo Kudo

Abstract This chapter first introduces the aim of this book and explains the book structure that provides chapters on theoretical discussions and practical applications of different framing in specific cases. Secondly, the chapter provides summaries of all previous chapters. Thirdly, the chapter describes that sustainability science has two main roles based on a premise that sustainable development as a trajectory-based concept, that are (i) examining the past patterns of trajectories that have brought societies to their present state, and (ii) designing the future based on the actions of the current generation. Reflecting these roles, the authors argue that sustainability science examines the intended and unintended consequences of actions taken by various actors. Lastly, the authors remark that an attitude to be flexible and accepting to other's framings is extremely important in order to have collaborative actions for sustainability.

Keywords Framing · Sustainability science · Intended consequence · Unintended consequence · Collaboration

T. Mino (✉)
Graduate Program in Sustainability Science-Global Leadership Initiative,
Graduate School of Frontier Sciences, The University of Tokyo, Kashiwa, Chiba, Japan

Department of Socio-Cultural Environmental Studies, Graduate School of Frontier Sciences,
The University of Tokyo, Kashiwa, Chiba, Japan
e-mail: mino@k.u-tokyo.ac.jp

S. Kudo
Graduate Program in Sustainability Science-Global Leadership Initiative,
Graduate School of Frontier Sciences, The University of Tokyo, Kashiwa, Chiba, Japan
e-mail: kudo@edu.k.u-tokyo.ac.jp

© The Author(s) 2020
T. Mino, S. Kudo (eds.), *Framing in Sustainability Science*,
Science for Sustainable Societies, https://doi.org/10.1007/978-981-13-9061-6_9

9.1 Chapter Summaries

This book aims at examining the applied framing in sustainability research. The editors try to achieve this goal in the first three chapters by focusing on the theoretical discussions on framing itself, and in the subsequent five chapters introducing different types of practical framing on specific topics or empirical cases. The present volume does not allow a full coverage of all types of sustainability issues; however, the editors believe the book serves as an initial step for reviewing different types of framing applied in sustainability research and actions.

Chapter 1 by Mino and Kudo describes what framing is in general and why discussing framing in sustainability science is essential when addressing sustainability issues. Framing explains how people perceive, understand, and interpret a particular topic or event based on the assumptions, social norms, and values that people have in their daily lives. Framing defines what situations are relevant to people, who should be responsible for them, and who should take measures to improve the situation or avoid possible undesirable situations. Sustainability is essentially a normative concept and it requires people to frame what to sustain in society. In reality, when addressing a sustainability issue, multiple framings by different actors always exist and often conflict to one another. Hence, acknowledging the multiplicity in framing is critically important in sustainability research. Sustainability experts are expected to facilitate the communication among multiple framings posed by different groups of people and lead the discussion that results in concrete actions for sustainability. In this chapter, Mino and Kudo propose a conceptual framework that suggests key elements to be examined when addressing sustainability issues. The framework combines holistic treatment and trans-boundary thinking to incorporate multiple framings when understanding the complexity of a sustainability issue. The proposed framework also contributes to verifying whether the proposed actions reflect both global sustainability manifestation and unique values based on individual, case-specific contexts.

Chapter 2 by Jerneck and Olsson discusses pluralism and unification in sustainability research. They argue that pluralism allows collaborations among scholars and social actors for sustainability, and that it also ensures a more theoretically informed understanding of society. Pluralism is a better way to share understanding about the complex sustainability challenges from the points of various social actors, hence it helps addressing complex sustainability challenges compared to conventional views such as environmental determinism, functionalism, and rational choice theory to be tackled. The authors suggest 'social fields and natural systems' as a new framing concept that integrates social science theory and a systems science perspective. This framing aims to bridge ontological barriers between natural science and social science, and avoids three common weaknesses of knowledge integration–namely, the use of environmental determinism, functionalism, and rational choice theory–to explain social changes. Additionally, a new theoretical typology was suggested to show how sustainability visions and pathways are linked with particular theoretical and methodological perspectives. The presented typology serves as a thinking tool for framing and reframing sustainability research.

Chapter 3 by Ness introduces four general approaches to frame sustainability challenges, which are DPSIR, causal look diagrams, multi-scale and multi-level perspective in transitions, and socio-ecological system framework. The chapter describes how these frameworks are taught at the international master's program in sustainability at Lund University (LUMES). The chapter provides reflections from teaching various framework approaches in sustainability science. The author suggests that multiple occasions to learn the approaches to frame complex sustainability challenges must be provided both within and outside of an educational program for students in order to obtain the key competencies required for becoming sustainability experts.

Chapter 4 by, Yokohari, Murakami, and Terada, introduces mixed patterns of land use by a case study of Tokyo. The authors argue that this mixed land use enhances the quality of living environment as well as resilience of cities by contributing to food security, especially at the time of natural disasters such as earthquakes. 'Value of grey' is the concept they use to describe such mixed pattern of land use in urban planning developed in Japan. Such strategy in land use has been developed in Japan in order to incorporate the particular condition of the country: frequent occurrence of natural disasters. This chapter argues that basic theories of modern urban planning initiated in Western Europe, where almost no threat of natural disasters is predicted, are not always applicable to Asian cities where natural disasters such as earthquakes, tsunami, and tropical typhoon disasters often cause serious damages.

Chapter 5 by Kudo discusses the meaning of rural sustainability in an aging and shrinking society. The chapter firstly reviews literature on past transition patterns of rural regions based on the multifunctionality discourse in rural studies. Then, the concept of placemaking, accompanied by a case study of *Monogatari* workshop conducted by the author, is introduced. This case study provides community-based perspectives about how a group of local youths collectively envision the future of their town. The chapter serves as an empirical study of framing at the communal scale. One framing tool that applies a retrospective view when envisioning the future state based on the present and the past of the town is introduced.

Chapter 6 by Onuki presents a case of Minamata disease, one of the most serious water-related pollution diseases that Japanese society has ever experienced. Because of its seriousness and its scientific as well as social complexity, multiple explanations on the cause of the incident were proposed by experts in environment-related disciplines. Unfortunately, it was these very multiple framings to the issue themselves that prevented prompt reactions to the possible causes of the disease and hindered relief measures to the affected people in the area. The case of Minamata disease provides an important lesson on the balance between the actions to the ongoing problems and the degree of emphasis on scientific investigation on the problems.

Chapter 7 by Esteban and colleagues presents two different types of coastal issues–namely tsunamis and sea level rise–seeking to examine the vulnerability and resilience of human life and society in the face of natural hazards through the reduction and management of risks. Their case studies examine the reconstruction of Otsuchi town in northeast Japan after the 2011 Tohoku Earthquake tsunami and small islands of the coast of Bohol following the 2013 Bohol Earthquake in the Philippines. Based on these cases, the chapter discusses how the time-scale through which a problem is scrutinized influences the framing of disaster risk reduction, and thus disaster preparedness and management.

Chapter 8 by Matsuda and co-authors elaborates on the current problems of food security as one of the most important development strategies to alleviate poverty. Relevant policies on food security and different ways to frame these policies are discussed. The unchanged frame of enhancing market economy is considered to be the main cause of decoupling of agricultural policies and nutrition policies. The consequence of such failure in the integration of market-based policies and the actual situation of agriculture on the ground has manifested itself in serious environmental degradation and poor health conditions.

9.2 Sustainability Science Examines Intended and Unintended Consequences of Framing

As summarized above, the chapters in this book provide various perspectives to elucidate sustainability issues such as contextual, spatial, and temporal, perspectives. By incorporating these, sustainability science raises scientific inquiries to explore sustainable pathways for societies. One premise in this approach is that society evolves in a path-dependent ways; hence, sustainability can be viewed as a trajectory-based concept. This perspective implies two main roles of sustainability science: (i) examining the past patterns of trajectories that have brought societies to their present state, and (ii) designing the future based on the actions of the current generation. In other words, the first role is providing a structural understanding on how the human society of today has ended up being faced with a wide range of sustainability issues; and the second role is advocating sustainability as the guiding principle to design future societies through collaborations among various actors. Framing, the primary focus of this volume, affects both these roles of sustainability science significantly on how the present state is being interpreted and how the future directions are envisioned.

Reflecting these two main roles, sustainability science plays a key role in examining the intended and unintended consequences of framing. Various measures to achieve sustainability are undertaken based on the visions of ideal or preferred conditions of society. When those originally envisioned conditions are realized, the outcomes are considered as intended consequences. However, unintended consequences also appear because of the specifying nature and unexpected effects of framing. Framing specifies what to be addressed and naturally prioritizes some issues over others. At the same time, framing may even slow down or prevent the envisioned conditions from being achieved because verifying what one framing brings is hardly possible prior to its application; hence, unintended consequences are often observed.

The Sustainable Development Goals (SDGs) scheme by the United Nations is one example of framing that is likely to have both intended and unintended consequences. SDGs consist of 17 goals with 169 targets covering a wide range of sustainability issues. These goals are to be achieved by 2030 under the slogan of "no one left behind." Since their introduction in 2015, countries in both developed and developing regions have adopted the scheme and are now conducting local initiatives to achieve the goals. Examining unintended consequences is equally important to all efforts made to achieve SDGs. For example, one potential unintended consequence of setting the 17 goals is the fragmentation of individual goals despite the fact that many of them are interlinked in reality. For instance, Goal 1, "No Poverty", is a comprehensive goal that must include hunger, education, access to clean water, and many other dimensions of sustainable development. In fact, all other 16 goals of SDGs are either directly and indirectly related to Goal 1.

Additionally, issues not included in the 17 goals may seem less important despite the fact that they have equal or even greater impacts on human society such as aging, mental health, and regional focus on rural areas. Because the topics related to the goals are highlighted by policies at all levels, insufficient consideration to other sustainability issues may be observed as one unintended consequence.

By presenting 17 goals clearly, the SDG scheme assumes that sustainable development can be achieved by filling these pieces of the Sustainable Development picture. Such perspective stands on a premise that the whole can be understood and fulfilled by understanding and integrating the parts. However, in reality, how the whole works cannot be understood merely by a compilation of understanding how parts function. This is because the interactions among the parts create generic functions that contribute to the capacity of the whole. One unintended consequence of setting individual goals in the SDGs scheme is the missing discussion on the interlinkages among the thematic goals. A holistic approach in sustainability science should be reflexive to the intended and unintended consequences of various efforts for sustainable development; this will be an important process when linking framing to actions for sustainability.

9.3 Concluding Remarks

The readers hopefully have learned from this volume what framing is theoretically, why it is relevant to sustainability, and how framing practically influences the actual implementation of countermeasures in overcoming the complex sustainability challenges. Framing is strongly influenced by past experiences, expertise and knowledge, social norms and values, and many other factors. Therefore, a difficulty in understanding different perspectives often rises when collaborating with actors from various fields even though such collaborations across different academic disciplines and sectors are often described as an essential approach when addressing sustainability issues. An attitude to be flexible and accepting to others' framings is extremely important to make collaborative actions successful especially in the context of sustainability. In any case, the collaboration of experts may provide opportunities to expose themselves to a wide range of new framings, improve their ability to understand the complexity of sustainability issues, and identify diverse approaches through policies, measures, and actions that can guide society towards more sustainable direction.

Name Index

© The Author(s) 2020
T. Mino, S. Kudo (eds.), *Framing in Sustainability Science*,
Science for Sustainable Societies, https://doi.org/10.1007/978-981-13-9061-6

Subject Index

A

Academia, 4, 5, 35, 142, 162
Accessibility
  food, 164
Acidification of the oceans, 146–148
Acidity, 147
Action plan development, 9
Actions, 4, 5, 9, 10, 12, 19, 26, 27, 30, 38,
    45, 50, 78, 80, 106, 109, 113, 121,
    122, 125, 127, 135, 138, 143, 155,
    176–180
Activism, 29
Actors, 5, 9, 24, 25, 35, 37, 45, 46, 101, 134,
    148, 160, 176, 178, 180
Adaptability, 60, 69, 92, 93
Adaptation measures, 134, 143
Adaptive management, 5, 21
Adulthood, 162
  stage, 162
Advantage, competitive, 19
Advocacy, 167
Aesthetic reasons, 141
Affordability, food, 164
Africa, 42, 147, 157, 160
Aged farmers, 77
Agential, 24
Aging population, *see* Aging society
Aging society, 7
Agricultural Land Act of 1952, 68
Agricultural policy, 102, 113, 114, 156, 168,
    178
Agricultural productivity, 156, 166
Agricultural Promotion Region (APR), 83
Agricultural Research Systems (NARS), 157

Aims, 4–8, 10, 12, 13, 18, 22, 25, 26, 37, 39,
    43, 51, 68, 82, 99, 100, 105, 114, 140,
    154, 155, 162, 166, 168, 176
Alliances, 167
Allotment gardens, 85–87, 89
Analysis, institutional, 25, 46
Anthropogenic causes, 41
Anti-modernity, 28–29
Approach
  comprehensive, 18, 50
  integrated, 18
  multimedia, 167
  multi-sectoral, 164
  participatory, 18, 37, 42, 167
  theoretical, 19, 28
  transdisciplinary, 164
Area division, 72, 81–83
Area division system (ADS), 81–83
Argenteuil, Ille de France, 61, 63
Arithmetic, 157
Asia, 58, 93, 157, 158
Asian cities, 58, 177
Assumptions, 8, 10, 18, 20–23, 25, 26, 40,
    109, 176
Asymmetric information, 166
Atoll islands, 143
Attitudes, 77, 166, 180
Attributes, 20, 47, 51, 105, 121
Authority, 125, 140
Availability, food, 42
Awareness, 28, 112, 113, 128, 163
  epistemological, 19
  ontological, 19
  theoretical, 19

© The Author(s) 2020
T. Mino, S. Kudo (eds.), *Framing in Sustainability Science*,
Science for Sustainable Societies, https://doi.org/10.1007/978-981-13-9061-6